Save the World with Code

About the Author

Lorraine Underwood has a degree in Computer Science and is a trained secondary school teacher. She currently teaches undergraduate students at Lancaster University in the UK. Lorraine is also a maker, educator, resource creator, super mum, and all around awesome person. She specializes in making coding fun and accessible for everyone. Her favorite medium to work in is shiny lights. Lorraine blogs about her projects on her website http://lorraine.mcunderwood.org/.

Save the World with Code

20 Fun Projects for All Ages Using Raspberry Pi, micro:bit, and Circuit Playground Express

Lorraine Underwood

New York   Chicago   San Francisco   Athens   London
Madrid   Mexico City   Milan   New Delhi
Singapore   Sydney   Toronto

Library of Congress Control Number: 2019039481

1 2 3 4 5 6 7 8 9   DSS   25 24 23 22 21 20

ISBN   978-1-260-45759-9
MHID   1-260-45759-1

Sponsoring Editor Lara Zoble	**Acquisitions Coordinator** Elizabeth Houde	**Indexer** Claire Splan
Editing Supervisor Donna M. Martone	**Project Manager** Patricia Wallenburg	**Art Director, Cover** Jeff Weeks
Production Supervisor Lynn M. Messina	**Proofreader** Judy Duguid	**Composition** TypeWriting

For the hubster, Phil,
my number one fan

Contents

PART ONE
Zombie Defense

PART TWO
Defend Your Home

PART THREE
Save the World

Preface

Welcome to my book. I wrote this book! Yes, me! My name is Lorraine, and I love coding. I think coding is the portal to everything—the portal to having fun, being creative, being serious, and yes, saving the world. Superheroes of the future will not have superpowers or a cape; they'll have a laptop and great Wi-Fi.

Acknowledgments

I am so grateful to be given the opportunity to write this book by McGraw-Hill, and I would particularly like to thank Elizabeth Houde for all her help and guidance, and Patty Wallenburg for her amazing editing.

I'm very lucky to live near a wonderful small school that my sons attend. The models in this book are all from my sons' school and are (and will be!) in the Code Club, which I run every week. Thanks to Indie, Olive, Rose, Ffion, Catherine, Jasper, and Bailey for patiently posing for my photos and for their endless smiles and giggles.

Thanks to my reviewers: Michael, Rachel, Hannah, Lynne, Phil, James, Carolyn, Liam, Angie, and Nicole.

Thanks to my sons, David and Sam—David for being a model in this book, testing out the code in the missions, and being more excited about this book than me and Sam for also being a model in this book, doing his best to help out, and constantly asking me when the book was ready.

Thanks to my Twitter friends, Laurence, Darren, and Paul, for their endless enthusiasm and inspiration. To Helen Leigh, for blazing the trail with her book *The Crafty Kids Guide to DIY Electronics: 20 Fun Projects for Makers, Crafters, and Everyone in Between* (McGraw-Hill Education TAB, 2018). To my lifelong friend Katie for keeping me going over long distance and my number two fan Rachel, who always knows just what to say.

Save the World with Code

Getting Started

With this book, you can code missions to help save the world using the BBC micro:bit, the Circuit Playground Express, and/or the Raspberry Pi. At the end of this Introduction, you will find a list of missions by device, difficulty, and whether you need extra equipment.

NOTE TO PARENTS/GUARDIANS
The Raspberry Pi is more suitable for older children or those with a lot of help from an adult, especially when setting it up. Many children will not have experienced Linux before, and the environment will not be familiar to them. The MakeCode platform for the BBC micro:bit and Circuit Playground Express is really easy to use. It is very similar to Scratch, which many children will have used before.

Start this Introduction by choosing your tool and learning your skills and how to get mission ready. Then move on to choosing a mission. Most missions can be completed individually; some missions are connected to previous missions. Some missions may refer to definitions or code from earlier missions. Jump right in and get started on saving the world!

Choose Your Tool: BBC micro:bit, Circuit Playground Express, or Raspberry Pi

BBC micro:bit

The BBC micro:bit is a small device packed with many features. You need a computer to code it, so it's not a full computer, but it is a microcomputer.

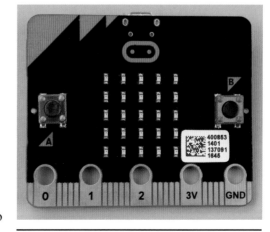

Figure I.1   The BBC micro:bit.

It was created by the British Broadcasting Company (BBC) and many other partners in the United Kingdom in 2015. In 2016, one million micro:bits were delivered for free to schools in the United Kingdom to inspire a generation of children to code. The micro:bit is now available to buy around the world, and many other countries have followed the United Kingdom's example of giving them out for free in schools to support the teaching of coding.

Some of the features available on the micro:bit that we're going to use in this book include:

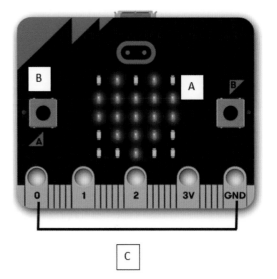

- Twenty-five light-emitting diodes (LEDs) on the front in a 5 × 5 grid [A]
- Two buttons [B]
- Touch and physical pins [C]
- Light sensor
- Temperature sensor
- Accelerometer
- Compass
- Radio

You can code LEDs to display text, numbers, and smiley faces! Besides displaying light, these LEDs can actually sense light too [A].

You can press button A, you can press button B, and you can press A and B together [B].

Figure I.2   BBC micro:bit up close.

The holes are called *pins*. You can add extra accessories to the micro:bit using these pins. They are touch sensitive [C].

Find out more about each feature on the micro:bit website: https://microbit.org/guide/features/.

The micro:bit is powered with a 2 × AAA battery pack or a USB socket plugged into a laptop. It's not recommended that you plug the micro:bit into a USB power bank or charger.

We're going to code the micro:bit using blocks called *MakeCode*, which was developed and is maintained by Microsoft. You can also code the micro:bit using microPython. Find out more about microPython at this website: http://docs.micropython.org/en/latest/.

Circuit Playground Express

The Circuit Playground Express was created by Adafruit as a successor to its popular Circuit Playground Classic. The Express has more sensors and is easier to program than the Classic. The Circuit Playground Express is a bit bigger than the BBC micro:bit. Some of the features on the Circuit Playground Express that we're going to use in this book include:

- Two buttons [A]

- Ten colored lights around the edge [B]
- Touch and physical pins [C]
- Light sensor
- Temperature sensor
- Microphone sensor
- Accelerometer
- Speaker

You can press button A, you can press button B, and you can press A and B together [A].

You can code these lights to show different colors. You can code all of them or just specific ones [B].

The holes are called *pins*. You can add extra accessories to the Circuit Playground Express using these pins. They are touch sensitive [C].

I power the Circuit Playground Express with the same battery pack as I use for the micro:bit: a 2 × AAA battery pack, but you can use a 3 × AAA battery pack and other types of batteries.

Find out more about each feature of the Circuit Playground Express on the Adafruit website: https://learn.adafruit.com/adafruit -circuit-playground-express.

We are going to code the Circuit Playground Express using Microsoft MakeCode as well.

Raspberry Pi

Unlike the BBC micro:bit and the Circuit Playground Express, the Raspberry Pi is a computer. Once you plug in a monitor, keyboard, and mouse, it's a fully working computer! But it has this huge difference over most computers: you can see its insides. You can add hardware to its inside and create and code a bigger and better computer. The Raspberry Pi doesn't have any built-in sensors like the BBC micro:bit or the Circuit Playground Express, but we're still going to build some very cool projects with it.

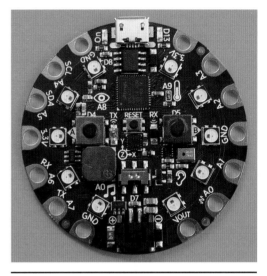

Figure I.3 The Circuit Playground Express.

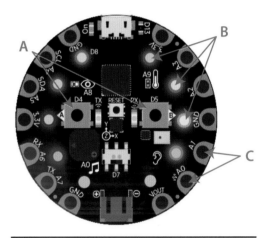

Figure I.4 Circuit Playground Express up close.

Figure I.5 Raspberry Pi.

There are many different types of Raspberry Pi units. In this book, we're going to be using the Raspberry Pi 3 Model A+, but all these missions will work on any other Raspberry Pi.

To get started with setting up your Raspberry Pi, follow this guide from the Raspberry Pi Foundation: https://www.raspberrypi.org/help/noobs-setup/2/.

Learn Your Skills: MakeCode or Python

Each device has multiple ways of being coded. We're going to code the micro:bit and Circuit Playground Express using MakeCode and a computer. And we're going to code the Raspberry Pi using Python, plugging the Raspberry Pi into a monitor with a keyboard and mouse attached.

For each mission, we'll look at MakeCode for micro:bit and then Circuit Playground Express. Then we'll move on to the Raspberry Pi.

Coding with MakeCode

MakeCode is a website that you can access via this address using a browser such as Google Chrome or Microsoft Internet Explorer: http://makecode.com.

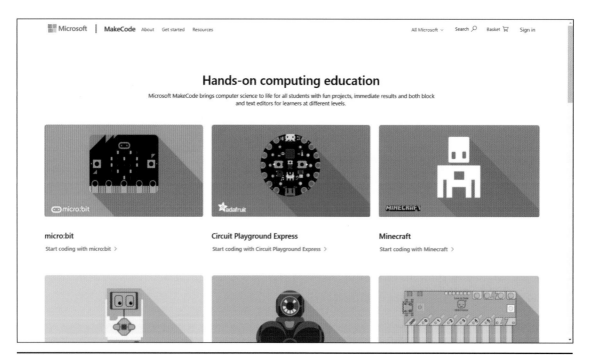

Figure I.6   MakeCode landing page.

From the MakeCode page you can select your device, the BBC micro:bit or the Circuit Playground Express. Next time you could go straight to the website for your device: https://makecode.microbit.org/ or https://makecode.adafruit.com/. The websites look slightly different from each other, but they act in pretty much the same way.

Projects

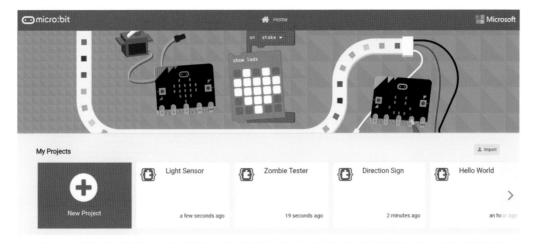

Figure I.7   My projects on the MakeCode micro:bit website.

Once you start creating multiple projects, they will appear on the MakeCode website. They are saved via your browser onto your computer. You won't be able to find them if you use a different computer. To start a new project, select **New Project**.

Figure I.8   New Project button.

MakeCode Platform

Let's have a look at the MakeCode website, on which we're going to code.

Figure I.9   MakeCode micro:bit coding screen.

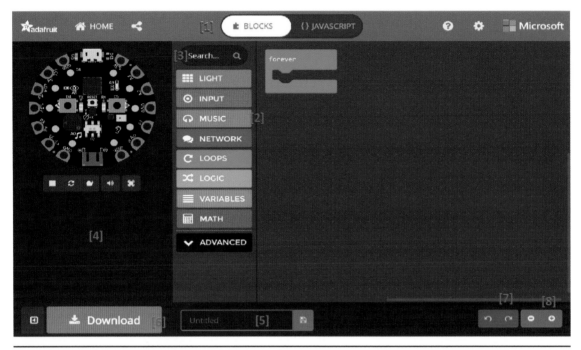

Figure I.10   MakeCode Circuit Playground Express coding screen.

You will see the following buttons:

- **Blocks/JavaScript** [1]. We're going to code in *blocks*. If you want to see what the code looks like in JavaScript, press this button to switch between the two.
- **Menu** [2]. These are the menus. When you click on one, you will find the blocks that make up your code
- **Search** [3]. If you can't remember the menu with the block inside, you can always use the Search field to type in the name of the block. I love this feature. I use it all the time!
- **Simulator** [4]. This is the simulator. This is super useful for testing your programs before you download them.
- **Project Name** [5]. Here you can name your projects. This is a really good idea so that you can keep track of which project is which.
- **Download** [6]. Guess what this button does?
- **Undo and Redo** [7]. The super-important Undo and Redo buttons.
- **Zoom** [8]. Zoom in and out.

Your First Program Using MakeCode

Using MakeCode is really good fun. You find the blocks you want, drag them onto your coding screen, and click them together. We're going to create a simple program to say "Hello World" when you start up the micro:bit and the Circuit Playground Express.

The *start* block should be on your screen when you first start, but if it's not:

1. In the micro:bit MakeCode, click on the **Basic** menu, and drag it out onto the screen.
2. In the Circuit Playground Express MakeCode, click on the **Loops** menu, and drag it out onto the screen.

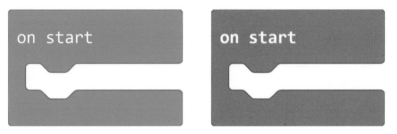

Figure I.11 *On start* block.

Then:

1. **micro:bit.** From **Basic**, drag out the block *show string "Hello!"*.
   a. Click on the word *hello* and type *Hello World*.
   b. **Circuit Playground Express.** From the **Light** menu, drag out the block *show ring*.
   c. Select your favorite color from the triangles in the middle of *show ring*.
   d. Select the circles around the edge of the circle to change them to your favorite color. Mine is green!
2. Place the block *show ring* inside the *on start* block. If your computer has sound, you should hear a satisfying click noise. Now when you move the *on start* block, the *show string/show ring* block should move with it like one large block made of two pieces.

Your program should now look like this:

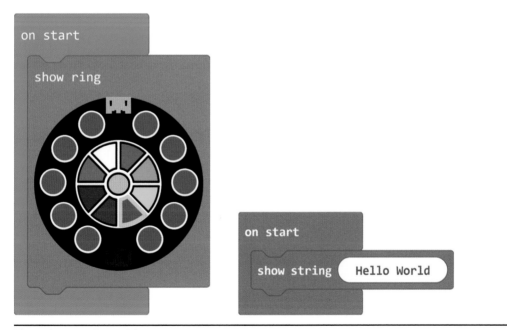

Figure I.12 Hello World.

You can add more blocks inside the ***on start*** block like so.

Delete Blocks

Try to keep your code nice and tidy. Don't let blocks overlap. Make sure that you can read it all clearly. Get rid of blocks that aren't doing anything.

Let's delete the ***forever*** block from our micro:bit code. It works the same way in Circuit Playground Express. Click and drag the block onto the menu like so:

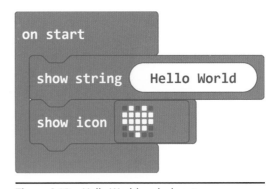

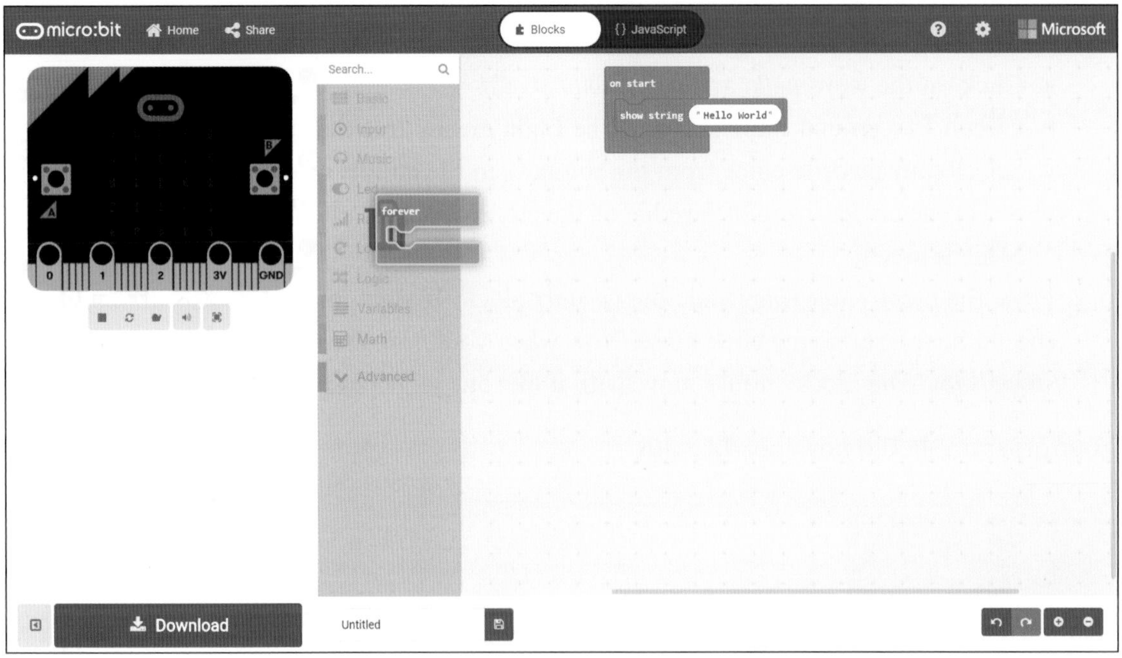

Figure I.13   Hello World and a heart.

Figure I.14   Deleting a block.

Be careful that you're only deleting one block. If blocks are attached to each other, they will all get deleted. Play around with the blocks in MakeCode to get used to them. Add some, delete some. Zoom in, zoom out. Undo, redo. Play!

Debug Code

Once you've finished with your code, you need to download the code to your device. It's a good idea to test it first. This can save you time when debugging. The simulator in MakeCode is really good for this. It lets you test your code on a virtual device.

This is what my micro:bit simulator looks like when it stops running. It scrolled the phrase *Hello World*; then it stopped on the love heart. What does yours show?

Figure I.15 micro:bit MakeCode simulator.

On the Circuit Playground Express, I can see that all the lights have changed to green.

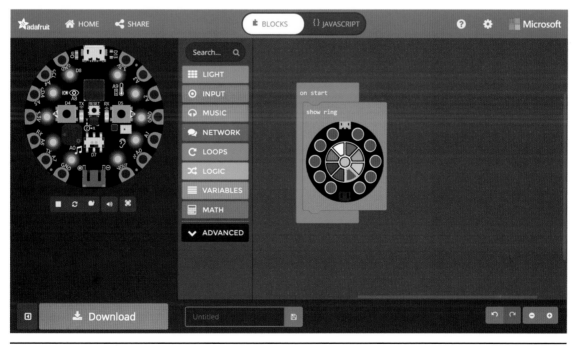

Figure I.16 Circuit Playground Express MakeCode simulator.

If your simulator didn't work the way you expected, then the real device isn't going to work either. Make sure that the code works fully on the simulator before downloading it.

Download Code

1. When you're ready, give your project a name that describes it and enter it in the bottom bar, next to the Download button.

Figure I.17   Name your project.

2. Press the Download button next to the project name. In Internet Explorer, a bar will pop up along the bottom of the screen and ask if you want to Open or Save the file.

Figure I.18   Internet Explorer pop-up.

a. Select **Save** to save the file to your Downloads folder.
b. Select **Open** to find the file.

In Chrome, the file will automatically download to your Downloads folder.

a. Select the small arrow next to the filename.
b. Select **Show in Folder** to find the file.

Figure I.19   Chrome download bar.

3. Both sets of steps should bring you to your Downloads folder. This is what it looks like on a Windows 10 machine.
   a. The micro:bit file will be called microbit-Hello-World.hex.
   b. The Circuit Playground Express file will be called circuitplayground-Hello-World.uf2.

Figure I.20   Downloads folder.

4. Plug the BBC micro:bit or Circuit Playground Express into your computer or laptop using a USB data cable. Some USB cables are for charging only. Make sure that your cable is a data cable.

5. Press the Reset button on the Circuit Playground Express.

6. Your device should appear on the left in the Downloads folder, called MICROBIT or CPLAYBOOT. You can see it in my Downloads folder (Figure 1.20).

7. Drag your file from the Downloads folder onto your device.

8. It should take just a few seconds for the file to transfer and then run. Adafruit has a guide and a video on how to do this for the Circuit Playground Express: https://learn. adafruit.com/adafruit-circuit-playground-express/downloading-and-flashing. A graphic showing how to do this for the BBC micro:bit is available on the micro:bit website. It also includes how to download the code using a Mac: https://microbit.org/guide/quick/.

Coding the Raspberry Pi with Python

The Raspberry Pi *is* a full computer. You plug in a monitor, keyboard, and mouse, and you're good to go! You also need an SD card with the operating system on it. All the missions in this book were created using Raspbian Stretch, which was downloaded in November 2018. The missions were tested again in April 2019 with the latest version.

With your Raspberry Pi plugged into a screen, keyboard, and mouse, select the Raspberry Pi logo in the top left-hand corner. This is your start menu. Select **Programming** and then **Python 3 (IDLE)**.

Figure I.21   Raspberry Pi start.

Python Shell

This opens a *shell*, a Python development environment. You can type code directly in here and see the results immediately. For example, type *"Hello World"* (including the quotes), and press ENTER on your keyboard.

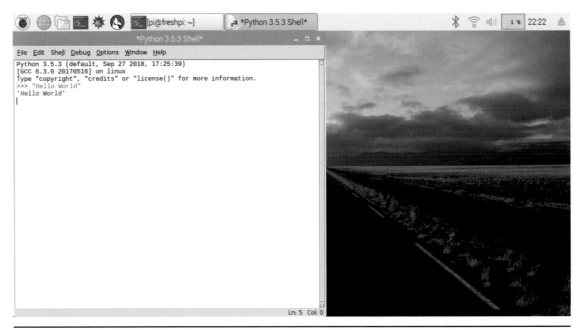

Figure I.22   Shell "Hello World."

Python Files

Our code isn't saved anywhere. When we close the shell, the code will go away. To create code that we can run again another day, select **File** and then **New File** from the shell window. A new window will open, and this is a Python file. The code won't run here. We'll need to type it here, save it, and then run it.

Let's try that now:

1. In the new empty file, type:

```
print ("Hello World")
```

2. Save the file by selecting **File** and then **Save** from the top menu. Call the file *hello*, and save it in the current folder, /home/pi.

Figure I.23   Save shell.

3. Now select **Run Module** from the top menu of *hello.py* and **Run Module F5.** Here we see the shell and the Python file opened together, just before I press **Run Module.**

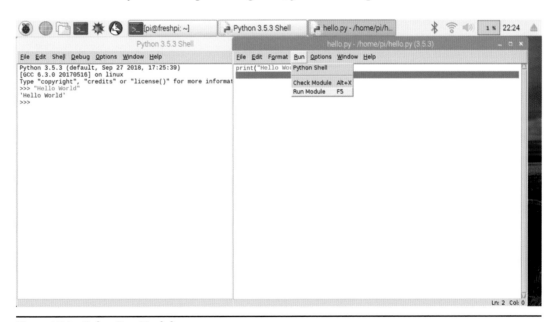

Figure I.24   Before **Run Module**.

4. When you press **Run Module**, the code will run in the shell. Your shell might be hidden behind your Python file. Drag the windows around to see them side by side.

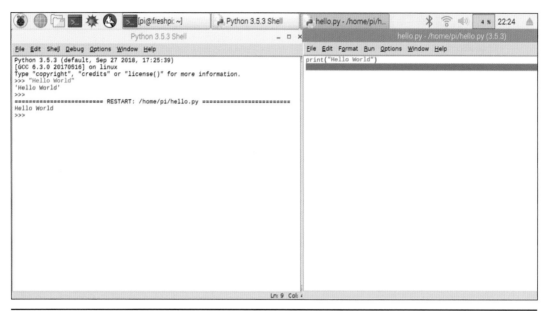

Figure I.25   After **Run**.

Unlike the blocks we use in MakeCode, Python is a text-based language. Every part of the sentence

```
print ("Hello world")
```

is important. **print** has to start with a small p. In Python 3, you have to have the parentheses around what you want to print. Your string started with quotation marks, and it has to end with them too. There are a lot of ways to fail in text-based languages. Yeh! But I will show you the way, young Padawan.

Terminal

Another way to run the code is to use the terminal. This will come into use later when we're downloading tools.

1. In the Python file, create a second file (select **File** and then **New**), and type this into the new file:

```
print ("Goodbye world")
```

2. Save it, and call it *goodbye*. Here are the two files *hello.py* and *goodbye.py* and the shell open together.

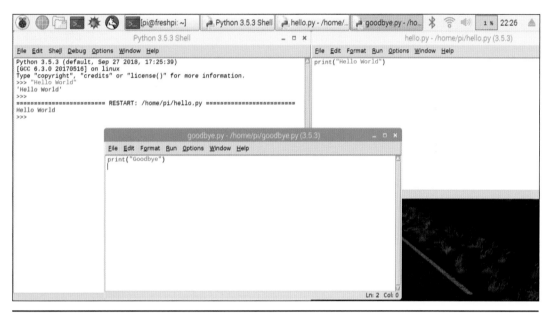

Figure I.26 Hello goodbye.

3. Open a new terminal window by selecting the terminal icon in the top toolbar.

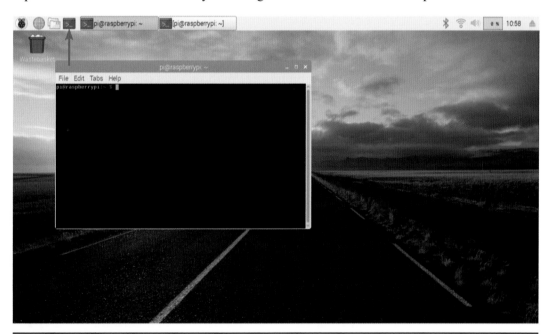

Figure I.27 Launch the terminal.

4. Type *python3 hello.py*, and press ENTER.
5. Type *python3 goodbye.py*, and press ENTER.

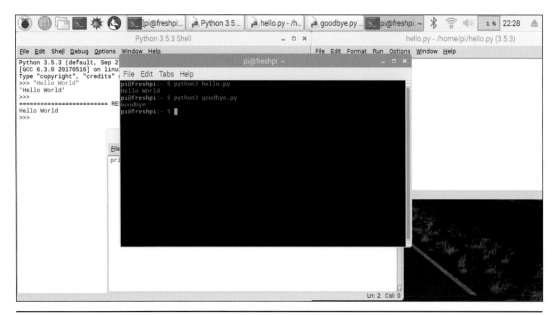

Figure I.28   Terminal.

The benefit of using the terminal is that you don't have to open the Python files to run them. You can actually create and edit Python files from the terminal, but it can be tricky to edit them.

To find out more about Python on the Raspberry Pi, check out the Raspberry Pi website: https://www.raspberrypi.org/documentation/usage/python/. You can write Python code in any way that suits you best. My missions will include the full Python code.

Choose Your Mission: How to Use This Book

You have your tools, and you know your skills. To create the missions in this book, we're going to run through five tasks for each mission. Let's use the *"Hello World"* example above. It's as easy as ABC . . . D and E!

Algorithm

Every mission will start with an algorithm. An *algorithm* is a plan, a set of step-by-step instructions to solve a problem. There can be several parts to an algorithm:

- **Sequence.** The order that steps need to be in.
- **Selection.** When a decision needs to be made.
- **Repetition.** When some steps need to be repeated.

Look at this algorithm for brushing your teeth:

```
If toothpaste is empty [LABEL: Selection]
     Buy more toothpaste
Else [LABEL: Selection]
     1. Put toothpaste on toothbrush. [LABEL STEPS 1-3 Sequence]
     2. Repeat until all teeth are clean: [LABEL Repetition]
          a. Brush a tooth.
     3. Rinse your toothbrush.
```

Every part of the algorithm is important. If the selection is wrong, you might be brushing your teeth with no toothpaste. If the sequence is out of order, you might end up putting toothpaste on the toothbrush for every tooth! If the repetition sentence is wrong, you might never stop brushing!

Algorithms are really important when it comes to coding. You should never start coding straightaway. You need a plan: an algorithm written in plain English is a good start.

Here's our algorithm for our first program:

```
When the device starts
     Display "hello world"
```

The next few steps are specific to the devices. If you only have a Circuit Playground Express, you will be able to skip the micro:bit build, code, debug, and expert level steps and go straight to the Circuit Playground Express section. And the same goes for micro:bit and Raspberry Pi.

Build

Some of the missions in this book require extra build steps, such as adding extra accessories or placing the device in a certain location. I'll describe these using detailed (and hilarious) photographs.

Some missions won't have any build steps, such as the Zombie Tester in Mission 1 for micro:bit and Circuit Playground Express. You just need to run the code, so the build section will just be a photograph.

The next sections will be repeated for each device for that mission. To find the missions for your device, see page 21 of Your Missions.

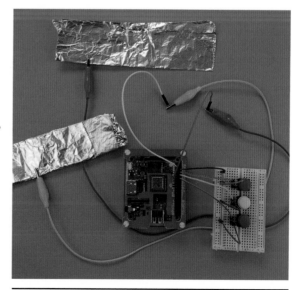

Figure I.29 Build photograph.

Code

The code will be broken down step by step until the final code is shown. Sometimes I'll add a bug for fun. But we'll fix it together. Here's the code for the micro:bit.

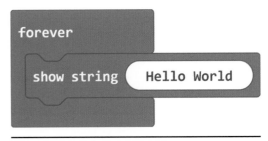

Figure I.30   Debugging.

WHAT'S A BUG?

A *bug* is a mistake in your code. It's something you didn't mean to happen. It's called a *bug* because American computer scientist Grace Hopper found a real bug, a moth, inside a computer in 1945 that caused the computer to malfunction.

Debug

After we've created our code using the algorithm, we might find a bug that we'll need to debug. The bug might be in the code, or it could be that our algorithm was wrong in the first place, and we need to add more code.

"Hello World" Debug

"Hello World" is displaying over and over again because it's a forever loop. We only wanted it to display once. Oops! Here's the correct code for the micro:bit.

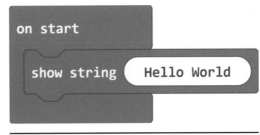

Figure I.31   "Hello World."

Expert Level

After we've coded and debugged the mission, I will challenge you to reach the expert level by asking you to add an extra feature to the mission. I won't give you the answer straightaway. All expert level answers are on the website accompanying this book (savetheworld.mcunderwood.org). Try the challenge first before looking for the answer. Push yourself with your new skills.

"Hello World" Challenge

Display your name when the user presses the A button on the BBC micro:bit. Make all the lights go blue when the user presses the B button on the Circuit Playground Express. Please see the website for the answer to this challenge (savetheworld.mcunderwood.org).

Kill Bugs in Your Code: How to Debug

Let's get this over with right now. You are going to have bugs in your code. Something will not work because you coded it wrong. It's okay! It happens. It happens to all of us.

In some of the missions, I'm even going to set you up by deliberately creating bugs. Fixing bugs will make you a really good coder. And spotting bugs is half the battle. So I'll be deliberately putting bugs in your code. See if you can spot them! I'll tell you, eventually. And there will be full bug-free code for all the missions in this book.

How to debug is all about how not to create bugs in the first place. Here are some tips to help you along the way:

1. **Keep it neat.** Don't overlap your blocks. Spread your blocks across the screen, and try to arrange them in an order that makes sense to you. Space out your Python code. While the code won't work in Python if it's in the wrong order, put all your libraries at the very top and then your variables. Order your functions in a way that makes sense.

2. **Get a buddy.** This is Duck; he is my duck. There are many like him, but he is mine.

   Explaining your code to someone else can really help you when debugging. There's even a method called *Rubber Ducking* where you explain your code to a rubber duck. Just talking through your code out loud can help you figure it out.

3. **Break it down.** If your code is quite long and complicated, break it down into smaller chunks. Try to test one piece at a time. Even remove code so that you can test smaller chunks. This seems like a bad idea, but sometimes it's hard to see a small bug in a lot of code.

Figure I.32   Duck.

Common Mistakes/Troubleshooting Your Build

Even the best programmers make mistakes. Here are some common mistakes I've made when using MakeCode and Python:

1. **Not connecting the blocks.** In MakeCode, the blocks need to be attached to work. If they are not attached, they won't be their normal color. The blocks will be gray. Check that all your code is attached and colorful.

2. **Not downloading your code.** Your code isn't doing what you want it to do. It's running the old code. In MakeCode, did you download the new code to your device? In Python,

did you save your changes? This is really common. Every time you make a change to your code, you have to download or save the code to the device again.

3. **Using the wrong cable.** You need a data cable to transfer the program from your computer to your device. There are some evil USB cables out there that are charge only. Try another cable to see if that's the problem.

4. **Low batteries.** If the micro:bit or Circuit Playground lights are looking a bit weak, you need to change the batteries.

5. **Loose cables.** Crocodile clips are always sliding off stuff on me. Wiggle the croc clips around to see if they're actually attached.

6. **Dodgy connections.** *Never* connect power to ground. Never ever, ever! If you do this, you risk draining your battery in one second flat or, worse, causing damage to your device. Make sure that the ground is connected to ground and power is connected to power.

7. **Too much power.** Make sure that you're using the right batteries for your device. Don't ever put a battery in that's more powerful than what's required, for example, three AAs instead of two AAs.

Next is a list of missions in the book by device and difficulty. Some missions need extra equipment, which is detailed next.

Your Missions

Part One: Zombie Defense

Mission	micro:bit	Circuit Playground Express	Raspberry Pi	Page No.
Mission 1: Zombie Detector	Level: Easy	Level: Easy	Level: Intermediate	27
Mission 2: Zombie Escape Sign	Level: Easy	Level: Easy	Level: Intermediate	37
Mission 3: Light Sensor to Detect When Vampire Zombies Are Awake	Level: Intermediate	Level: Easy		51
Mission 4: Light-Up Backpack	Level: Easy	Level: Easy		57
Mission 5: Light-Up Attack Sword for Battling Vampire Zombies	Level: Intermediate	Level: Intermediate		75
Mission 6: Reaction Game to Test for Zombies	optional Level: Advanced		optional Level: Intermediate	91

Extra equipment.

Part Two: Defend Your Home

Mission	\multicolumn{3}{c}{Device}	Page No.
	micro:bit	Circuit Playground Express	Raspberry Pi	
Mission 7: Name Badge to Identify Valid Family Members	Level: Easy	Level: Easy		105
Mission 8: Door Sensor to Tell When Your Room Is Under Attack	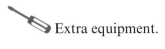 Level: Intermediate	Level: Intermediate	Level: Intermediate	109
Mission 9: A Lock to Protect Your Sword	Level: Advanced	Level: Advanced		123
Mission 10: Cookie Jar Protector	Level: Advanced	Level: Advanced		133
Mission 11: Number Lock for Your Devices	Level: Advanced	Level: Advanced	Level: Advanced	143
Mission 12: Mobile Alarm for Your Devices	Level: Advanced	Level: Advanced	Level: Intermediate	167
Mission 13: Floor Mat Alarm		Level: Easy	Level: Intermediate	177
Mission 14: Treasure Box Alarm	Level: Intermediate	Level: Intermediate		185

 Extra equipment.

Part Three: Save the World

Mission	micro:bit	Circuit Playground Express	Raspberry Pi	Page No.
	Device			
Mission 15: Step Counter				195
	Level: Easy	Level: Easy		
Mission 16: Bike Indicator				201
	Level: Easy	Level: Easy		
Mission 17: Moisture Sensor for Your Plants				211
	Level: Easy	Level: Easy		
Mission 18: Temperature Monitor	Level: Easy	Level: Easy	Level: Advanced	219
Mission 19: Temperature Alarm	Level: Advanced	Level: Advanced	Level: Advanced	235
Mission 20: Window Alarm				237
	Level: Advanced	Level: Advanced	Level: Advanced	

Extra equipment.

PART ONE
Zombie Defense

The first path to saving the world is defending yourself from a zombie (or other monster) invasion. These projects have detailed code instructions and thus are perfect for your first missions.

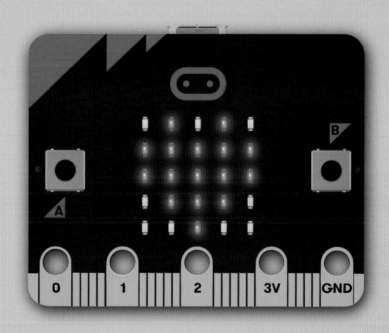

Zombie Detector

Sometimes it's difficult to tell whether someone is a zombie. Pale skin and a slow limp are obvious signs, but maybe the person is just not feeling well, or he or she has stubbed his or her toe? Follow this project to create an electronic test to see whether your family members are zombies or not.

To protect ourselves from zombies, we're going to create a touch sensor because it's a well-known fact that zombies don't conduct electricity like humans do.

Algorithm

Let's figure out what it is we're doing in simple English:

```
When device is touched, then:
    Person is human. Show a tick/green light
```

This seems easy enough! Let's get building.

micro:bit

Build

The micro:bit can detect a conductive item (you) between a data pin and ground. To test the suspected zombie, ask him or her to hold the GND pin on the right of the micro:bit and the pin labeled 0 on the left. To get the tick, the suspected zombie has to let go of 0.

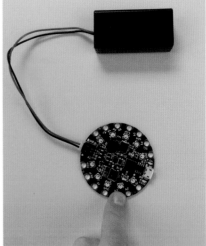

Figure 1.1   Zombie!

Code

Go to the MakeCode website for the micro:bit (https://makecode.microbit.org/), and select **New Project**.

1. Let's keep things tidy: delete the blocks *on start* and *forever*.
2. From the **Input** menu, drag out *on pin P0 pressed*.
3. From the **Basic** menu, drag out *show icon*, and drop it inside *on pin P0 pressed*.
4. Change the love heart icon to a tick icon by clicking on it and selecting the tick.

    This block acts like a button pressed. A button is pressed when it is pressed down and then let go. To test with the micro:bit, you will need to get the suspected zombie to press and then let go of pin 0. The suspected zombie can keep hold of the GND pin the whole time.
5. At the bottom of the screen, instead of Untitled, give your project a name.
6. Download the code to the micro:bit, and test it out. (See the Introduction.)

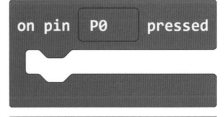

Figure 1.2   Pin pressed.

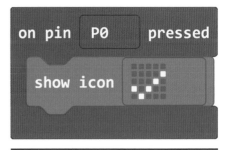

Figure 1.3   Tick.

Debug

Now, here's the problem: this code only works once, for one person. This mission is to test the whole family. We're going to have a queue of suspicious people whom we will want to test!

Let's clear the tick after a few seconds so that we can test again.

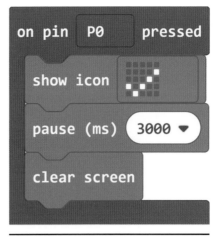

1. From the **Basic** menu, drag out *pause (ms) 100*, and place it under the *show icon.*
2. Change *100* to *3000*. This is 3 seconds. (See the box "There Are 1,000 Milliseconds in a Second.")
3. From the **Basic** menu, click on the submenu **More**.
4. Drag out *clear screen*.
5. Add it under the *pause (ms) 3000* block.

Now we can test more than one suspected zombie.

Figure 1.4 Debug.

THERE ARE 1,000 MILLISECONDS IN A SECOND

There are 1,000 milliseconds in a second. Four seconds is 4,000 milliseconds. How many milliseconds in half a second?

Expert Level

Here's a challenge for you to level up your skills. Until a human presses a button, show a cross. We're changing a lot about the project. Let's create a new algorithm:

```
On Start
    Show a Cross
When Pin is touched then
    Person is human! Show a Tick
    Pause for 3 seconds
    Clear the screen
    Show a Cross
```

Answers are found on this book's website (savetheworld.mcunderwood.org)! Don't peek. Try this challenge yourself.

Circuit Playground Express

Build

The Circuit Playground Express can detect when you touch some of its pins, such as A2.

Code

Go to the MakeCode website for Circuit Playground Express (http://adafruit.makecode.com), and select New Project to start coding. On the Circuit Playground Express, the touch-sensitive pins are A1 to A7.

1. Let's keep things tidy: delete the *forever* block.
2. From the **Input** menu, drag out *on button A click*.
3. Change *button A* to *pin A2*.
4. Change *click* to *down*.
5. From the **Light** menu, drag out *set all pixels to*, and place it inside *on pin A2 click*.
6. Change the red circle to a green one.
       Time to test your suspected zombie!
7. At the bottom of the screen, instead of Untitled, give your project a name.
8. Download the code to the Circuit Playground Express, and test it out. (See the Introduction.)

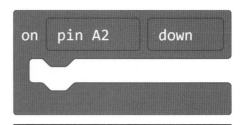

Figure 1.5   On pin.

Figure 1.6   Set pixels.

Debug

When testing my zombie sensor, I noticed that the green light stays on all the time. How do we know the suspected zombie has touched the pin? Let's clear the lights after a few seconds so that we can test again.

1. Add a *pause 100 ms* block from the **Loops** menu.
2. Change *100* to *3000*. This is 3 seconds. (See the box "There Are 1,000 Milliseconds in a Second.")
3. From the **Light** menu, scroll to the very bottom for the *clear* block.
4. Add this under *pause 3000 ms*.

Time to test *all* the suspected zombies.

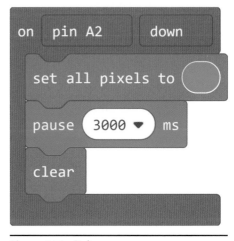

Figure 1.7   Debug.

Expert Level

Here's a challenge for you to level up your skills. Until a human presses a button, show red lights. We're changing a lot about the project, so let's create a new algorithm:

```
On Start
    Show all red lights
When Pin is touched then
    Person is human! Show all green lights
    Pause for 3 seconds
    Clear the lights
    Show all red lights
```

Answers are on this book's website (savetheworld.mcunderwood.org). Don't peek. Try this challenge yourself.

Raspberry Pi

Build

None of the pins on the Raspberry Pi are touch sensitive. You need to add an extra piece of equipment to run this project on a Pi.

I used Pimoroni's Touch pHAT. This is a cheap add-on board that sits on the Raspberry Pi pins and creates six capacitive touch buttons. You and/or your adult will need to solder a female header to the Touch pHAT to get it working with the Raspberry Pi. Pimoroni has some guides on

Figure 1.8   Raspberry Pi Touch pHAT.

soldering on its website (https://learn.pimoroni.com/tutorial/sandyj/soldering-phats). *Note:* When you add the hat to the Raspberry Pi, always turn the Pi off first.

Code

Download the Library

1. Let's start by downloading some libraries.

> ### DOWNLOAD A LIBRARY
> The Raspberry Pi doesn't have all the Python code in the world installed and ready to go. Sometimes you have to go and download it from the internet.

   Your Raspberry Pi needs to be connected to the internet to download the library. Install the Touch pHAT library from Pimoroni first.

2. Open up a terminal by selecting the terminal icon on the Raspberry Pi home screen.

3. Type:

```
curl https://get.pimoroni.com/touchphat | bash
```

   and press ENTER. The line just before the word *bash* is tricky to find on your keyboard. On my keyboard, it's to the left of the letter *Z*. You need to press SHIFT and that key to get that line to appear.

4. I had to type *Y* and press ENTER twice to continue—once to turn on I2C, which is turned off by default, and again to install the Pimoroni examples and documentation.

Figure 1.9 Terminal window.

5. When step 4 has finished, in the terminal again type:

```
pip3 install savetheworld
```

and press ENTER.

Figure 1.10 Installing the second library.

This will install my library for this mission.

Code the Pi

1. From the Raspberry Pi menu, select **Programming** and then **Python 3**.
2. Select **File** and then **New** to open up a new file to type code in.
3. Let's start by importing some libraries.

LIBRARY
Python doesn't have all the code it needs loaded all the time. Sometimes we need to tell Python to go fetch the code we need. We do this by typing "Import" and the name of the library. Earlier, we were downloading the libraries. Now we're importing them into our code.

The library *savetheworld touch* has a function in it called **touched**. We can send this function the name of the button we want to know about. This function returns either True (the button was pressed) or False (the button was not pressed). We can also send the word *Any* to check on any of the buttons on the Touch pHAT.

```
1  import time
2  from savetheworld import touch
```

Figure 1.11   Code.

1. Let's see what this returns. Run the program by selecting F5 on your keyboard or by clicking **Run** and then **Run Module** from the top menu.

```
1  import time
2  from savetheworld import touch
3
4  print(touch.touched("Any"))
```

Figure 1.12   Code.

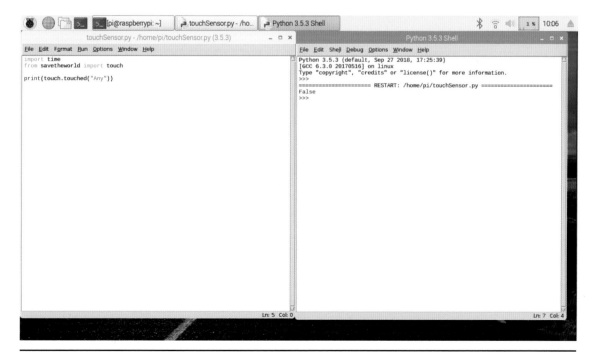

Figure 1.13   Running function.

2. Did you manage to get the program to display True? I didn't! This code runs once and only once. Let's put it in a forever loop and try again. You need to indent the last sentence so that it runs inside the while loop. Also watch out for the colon!

```
1  import time
2  from savetheworld import touch
3
4  while True:
5      print(touch.touched("Any"))
```

Figure 1.14   Code.

3. Run this code, and press a button on the Touch pHAT.

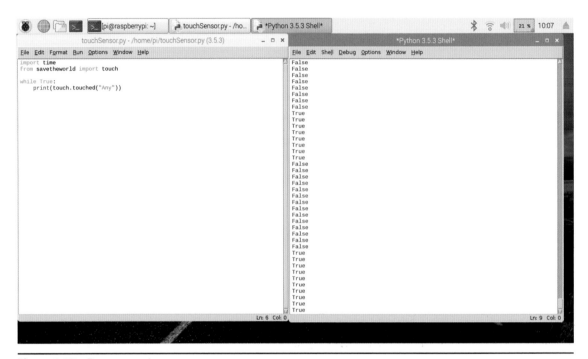

Figure 1.15   Forever running.

Success! The program spotted that I pressed some buttons on the Touch pHAT.

1. Click on the Shell that's displaying True and False, and press CTRL and C on the keyboard to stop the program from running. This will interrupt the program and show an error—but that's fine for now.

   We want to know when a button is pressed. We want to know when this function returns True.

2. I've put what the function returns into a variable called *phatStatus* so that I can check phat status in my if statement.

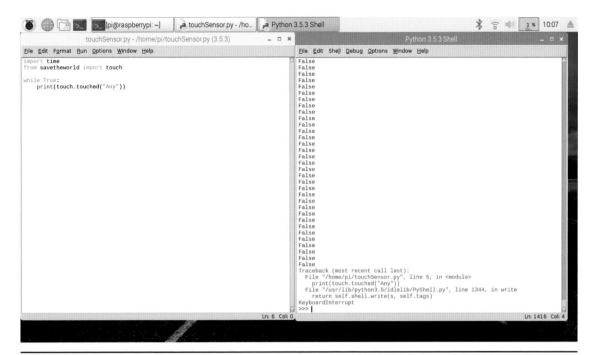

Figure 1.16 Force stop.

VARIABLES

A *variable* is somewhere we can store information. A great example of a variable is your score in a game. The score can be increased and decreased, we can see the score, and if you're a big cheat, you can change the score altogether. The same goes with variables! Variables can also be different data types such as text and numbers.

There are several ways you could write this code. You could leave out *phatStatus* altogether and use:

```
if touch.touched("Any"):
```

The line ***if phatStatus:*** could be rewritten as ***if phatStatus is True:***. I would suggest that you, as a beginner, use whatever makes the most sense to you.

Now you have a Raspberry Pi Zombie Detector. What message will you display instead of "Button pressed"?

```
1  import time
2  from savetheworld import touch
3
4  while True:
5      phatStatus = touch.touched("Any")
6      if phatStatus:
7          print ("Button pressed")
```

Figure 1.17 Checking the variable.

You could direct the zombie to only press button A; see if the zombie can follow instructions. Then use the code

```
touch.touched("A")
```

to check only the A button.

Debug

When you run the preceding code, it does work, but I get a lot of messages at once from a single button press. I think this isn't how the program should work.

Put in a small pause after printing the message "button pressed." This is just enough time for me to press my finger down and let go for a correct zombie test.

```
1  import time
2  from savetheworld import touch
3
4  while True:
5      phatStatus = touch.touched("Any")
6      if phatStatus:
7          print ("Button pressed")
8          time.sleep(0.5)
```

Figure 1.18   Pause.

Expert Level

Here's a challenge for you to level up your skills. Until a human presses a button, show a cross. If the suspect is a human, show a tick. Then clear the screen for the next suspect. We're changing a lot about the project, so let's create a new algorithm:

```
On Start
    Show a cross
If Pin is touched then
    Person is human! Show a tick
    Pause for 3 seconds
    Clear the screen
    Show a cross
```

Hint: Use the code from Mission 2 to draw graphics in Python. See the answer on this book's website (savetheworld.mcunderwood.org).

Zombie Escape Sign

When running away from zombies, it's easy to lose your friends. Create this direction sign to make sure that your friends know which way to go. We want to be able to set which arrow to display using the buttons on the micro:bit and the Circuit Playground Express. For the Raspberry Pi, we're going to use the letters *L* and *R* on the keyboard for left and right.

Algorithm

```
On button A/L press
    Show a left arrow
On button B/R press
    Show a right arrow
```

micro:bit

Build

When you've finished coding, unplug the USB cable, and plug in a battery pack. Place your micro:bit in a location where there are two directions: left and right.

Code

1. From the **Input** menu, drag out *on button A pressed*.

Figure 2.1 micro:bit direction sign in place.

2. From the **Basic** menu, drag out *show leds*, and place it inside *on button A pressed*.
3. Click on the squares in *show leds* to make a left arrow like the picture.
4. From the **Input** menu, drag out *on button A pressed*.
5. Change *A* to *B*.
6. Click on the squares in *show leds* to make a right arrow like the picture.

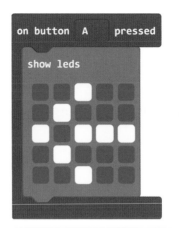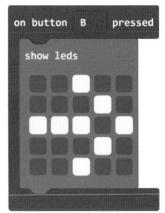

Figure 2.2   Arrow code.

Top Tip: Instead of creating the code from steps 4 to 6 again, right-click on the **on button A click** block and select *duplicate*. Then change *A* to *B*, and color the lights in.

Debug

Okay, slight problem. The zombies can see the direction sign. They'll just follow you, catch you, and eat you. Why not clear the lights after 3 seconds so that your friends will see them but any slow-moving zombie won't?

Our new algorithm is:

```
On button A press
    Show a left arrow
    Pause for 3 seconds
    Clear screen
On button B press
    Show a right arrow
    Pause for 3 seconds
    Clear screen
```

Do you remember how many milliseconds in a second? Look up the first mission in this section to find out. Also remember that the *clear screen* block is under the **Basic** menu; then you need to click the **More** submenu.

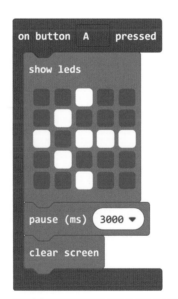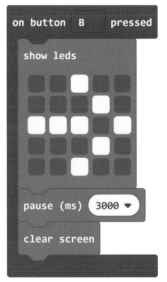

Figure 2.3   Debugged.

Expert Level

Use the light strip from Mission 4 to make a bigger, brighter direction sign.

Circuit Playground Express

Build

When you've finished coding the Circuit Playground Express, unplug the USB cable, and plug in a battery pack. Place your Circuit Playground Express in a location where there are two directions: left and right.

Code

Let's use the lights on the Circuit Playground Express as arrows. My lights will turn green, but you could choose any color you want.

1. From the **Input** menu, drag out *on button A click*.
2. From the **Light** menu, drag out *show ring.*
3. Click on the green triangle, and color all the lights on the left green by clicking on the circles.
4. Click on the gray dot in the middle of *show ring*, and turn off all the lights on the right by clicking on the circles.
5. From the **Input** menu, drag out *on button A click*.
6. Change *A* to *B*.
7. From the **Light** menu, drag out *show ring*.
8. Click on the green triangle, and color all the lights on the right green by clicking on the circles.
9. Click on the gray dot in the middle of *show ring*, and turn off all the lights on the left by clicking on the circles.

Figure 2.4   Arrow code.

Top Tip: Instead of creating the code from steps 4 to 7 again, right-click on the **on button A click** block, and select *duplicate*. Then change *A* to *B* and color the lights.

Debug

When testing, I pretended to be a zombie and chased after my sons. I could see the direction sign and so I knew which way they went! Let's clear the lights after 3 seconds so that your friends will see them but any slow-moving zombie won't.

Our new algorithm is:

```
On button A press
    Show a left arrow
    Pause for 3 seconds
    Clear screen
On button B press
    Show a right arrow
    Pause for 3 seconds
    Clear screen
```

Do you remember how many milliseconds are in a second? Look up the first mission in this section to find out. Also remember that the *pause* block is under the **Loops** menu.

Figure 2.5   Debugged.

Expert Level

In Mission 4, I show you how to add a light strip to a backpack. You could add this light strip here to make a bigger, brighter direction sign.

Raspberry Pi

Build

Not sure how happy your parents are going to be to find the TV in the middle of the hallway? There are some small screens you can buy for the Raspberry Pi and even some really cool LED grids. I'm just going to do it! Here is the code for displaying arrows on a TV screen.

Code

We're going to create one file for the Raspberry Pi that will display an arrow depending on which key you press on the keyboard.

Figure 2.6   TV in place.

Let's figure out how to draw an arrow first.

1. We're going to use the turtle library, so let's start with importing that. Learn what a library is in Mission 1.

> ### TURTLE LIBRARY
>
> In Python, the turtle library gives you access to a little turtle that you can code to move around the screen like a pen. You can change the color of the turtle, change its direction, and even lift the pen up to stop drawing as you move.

2. Create a variable called *myArrow*, and set it to be our turtle. (Learn more about variables in Mission 1.)

3. Color the turtle blue, and set the pen **size**. Setting **20** is quite thick, but I want my arrow to be really big on my screen. If your screen is smaller, you can change this number.

```
1  import turtle
2
3  myArrow = turtle.Turtle()
4  myArrow.color("blue")
5  myArrow.pensize(20)
6  myArrow.forward(200)
```

Figure 2.7   Code.

4. Forward! We have to set a distance we want to go. I'm going to start with 200. You might want to change this depending on the size of your screen.

5. Run the code by pressing F5 on your keyboard or by selecting **Run** and then **Run Module** from the top menu.

Figure 2.8   Forward!

6. Oooh, nice. By now, we've written a lot of code. At this stage, I like to add comments so that I know what I'm doing.

> ## COMMENTS ARE NOT CODE
>
> In Python, a comment starts with #. Comments are like a Post-it note about what your code is doing. They're vital to coding, and you should use them all the time.

7. Now we need to turn a certain number of degrees to draw the arrow—120 to be exact.

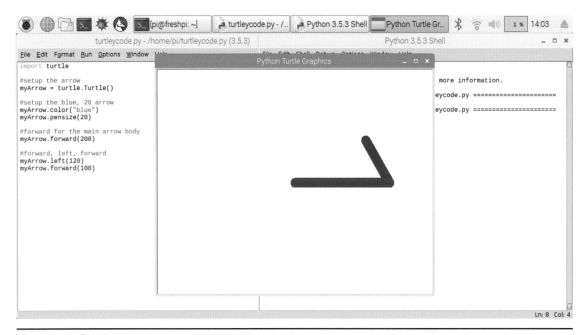

Figure 2.9   First turn.

8. Let's go backward, turn 240 degrees, and draw the rest of the arrow.

```python
import turtle

#setup the arrow
myArrow = turtle.Turtle()

#setup the blue, 20 arrow
myArrow.color("blue")
myArrow.pensize(20)

#forward for the main arrow body
myArrow.forward(200)

#forward, left, forward
myArrow.left(120)
myArrow.forward(100)

#turn around and draw the third line
myArrow.backward(100)
myArrow.right(240)
myArrow.forward(100)
```

Figure 2.10   Code.

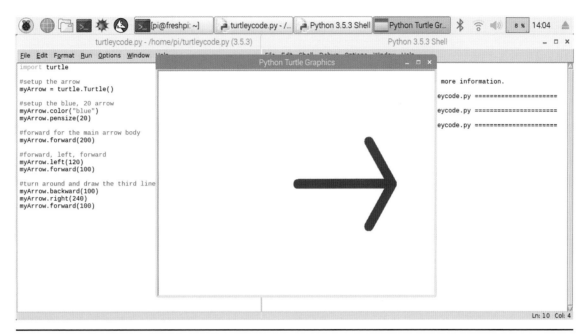

Figure 2.11 An arrow!

Ta-da! Beautiful. This arrow worked because at the start of our program we were facing in the right direction, and we were at location 0, 0. If we were to copy and run the code again, it wouldn't work.

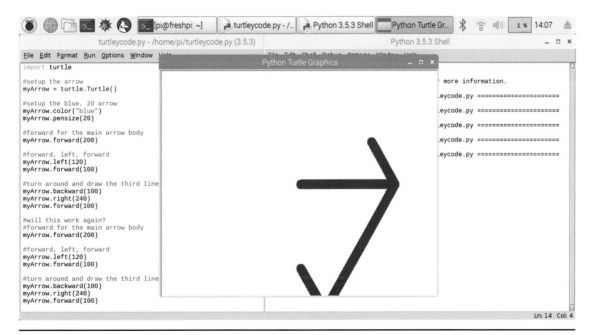

Figure 2.12 Wrong way.

At the start of each arrow, we need to set its position and direction. To separate right from left, we're going to separate the code out into functions.

Functions

A *function* is a piece of code that is separate from the main code. Functions have names, and you run the code inside a function by calling its name. You can send information to a function, and it can send information back.

In Python, a function starts with **def** and the function name. It has parentheses and then a semicolon. Everything inside the function is indented. Once you've finished the function, move back to the start of the sentence.

```
def functionName(dataIn):
    Code in the function
    Code in the function
    return dataOut
Code not in the function
```

Then you call the function, putting the data into a variable.

```
DataIWant = functionName(dataISen
dTheFunction)
```

It's a good idea to set up all your functions at the start of your code.

1. Let's create the right arrow function, setting the position and heading. We're not sending any data to the function or even getting any data back. This function will just draw an arrow on the screen.

   If you try to run the code, you'll notice that nothing happens. You have to call a function in order for the code to run.

2. Call the function by using its name.

```
1  import turtle
2
3  #setup the arrow
4  myArrow = turtle.Turtle()
5
6  def rightArrow():
7      #setup the blue, 20 arrow
8      myArrow.color("blue")
9      myArrow.pensize(20)
10
11     #set its position and direction
12     myArrow.penup()
13     myArrow.setpos(0,0)
14     myArrow.setheading(0)
15     myArrow.pendown()
16
17     #forward for the main arrow body
18     myArrow.forward(200)
19
20     #forward, left, forward
21     myArrow.left(120)
22     myArrow.forward(100)
23
24     #turn around and draw the third line
25     myArrow.backward(100)
26     myArrow.right(240)
27     myArrow.forward(100)
```

Figure 2.13   Code.

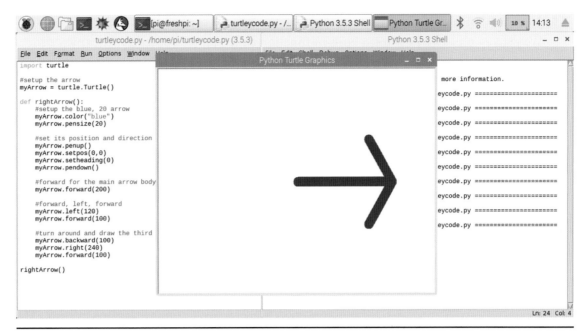

Figure 2.14 Function called.

3. Let's create the *leftArrow* function. It's pretty much the same as the *rightArrow* function. We're just facing a different way, and I want my left arrow to be a different color.

4. Let's call them one after the other to check whether they work.

```
1  leftArrow()
2  rightArrow()
```

Figure 2.15 Code.

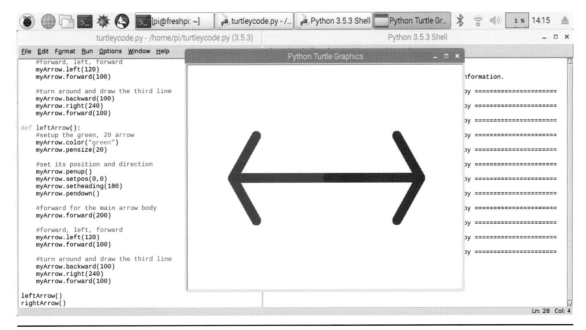

Figure 2.16 Left and right.

5. Now that we've got our two functions working, let's get the key from the keyboard. We get key presses by calling a built-in function called *input*. We send the text we want to display to the screen by placing it between the parentheses. It returns what the user types. We're going to capture what the *input* function returns into the variable *userEntry*.

```
#get a letter from the user
userEntry = input ("Enter your direction, L or R: ")
```

Variable that captures what the function returns | The function name | Data we are sending the function

Figure 2.17   userEntry.

6. Now let's check whether the key is an *L* or an *R* using **selection**.

7. Run the program. You might have to move the turtle screen off the shell screen. You have to click on the shell screen to type your text.

```
1  #get a letter from the user
2  userEntry = input("Enter your direction, L or R: ")
3
4  #L = call the leftArrow function
5  if userEntry == "L":
6      leftArrow()
7  #R = call the rightArrow function
8  if userEntry == "R":
9      rightArrow()
```

Figure 2.18   Code.

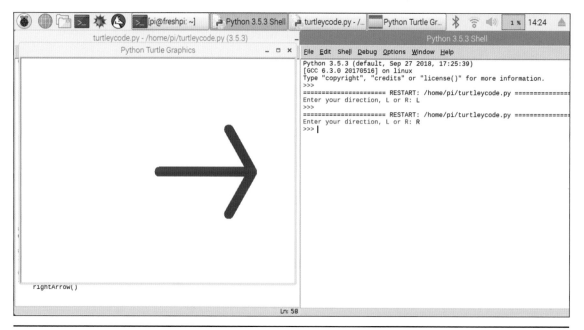

Figure 2.19   User entered R.

Every time you run the program, you can enter *L* or *R* and it will draw the arrow!

Debug

Once the arrow is on the screen, it stays on the screen. Zombies, monsters, and even unwelcome visitors can all see it! Let's clear the arrow after 3 seconds so that any slow-moving zombies, monsters, or relatives won't see it.

Our new algorithm is:

```
On button L press
    Show a left arrow
    Pause for 3 seconds
    Clear screen
On button R press
    Show a right arrow
    Pause for 3 seconds
    Clear screen
```

1. To pause for 3 seconds, we need to import the time library. Keep all your libraries in the same place, at the top of the file.

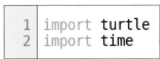

Figure 2.20 Code.

Tip: If you don't import the time library, you'll get this error—just like me!

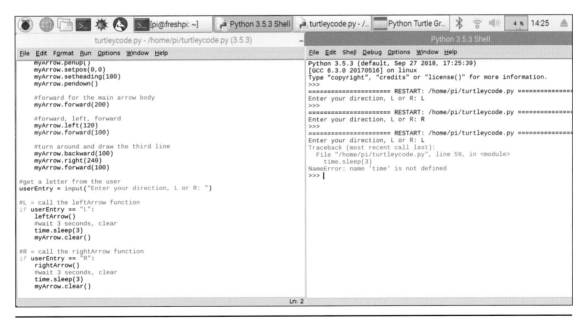

Figure 2.21 ERROR!

2. Here's the code to pause for 3 seconds and clear the arrow.

3. Now let's put everything in a forever loop so that it keeps asking you for the arrow as you run round and round in circles (see next page for the full code).

Expert Level

Try out your new direction sign. The turtle library is a really fun library.

```
1  while True:
2      #get a letter from the user
3      userEntry = input("Enter your direction, L or R: ")
4
5      #L = call the leftArrow function
6      if userEntry == "L":
7          leftArrow()
8          #wait 3 seconds, clear
9          time.sleep(3)
10         myArrow.clear()
11
12     #R = call the rightArrow function
13     if userEntry == "R":
14         rightArrow()
15         #wait 3 seconds, clear
16         time.sleep(3)
17         myArrow.clear()
```

Figure 2.22   Code.

Can you draw shapes with it? Try creating a new function called *square* that draws a blue square. Use the turtle library website to find other colors (https://docs.python.org/3.3/library/turtle.html).

```python
1  import turtle
2  import time
3
4  #setup the arrow
5  myArrow = turtle.Turtle()
6
7  def rightArrow():
8      #setup the blue, 20 arrow
9      myArrow.color("blue")
10     myArrow.pensize(20)
11
12     #set its position and direction
13     myArrow.penup()
14     myArrow.setpos(0,0)
15     myArrow.setheading(0)
16     myArrow.pendown()
17
18     #forward for the main arrow body
19     myArrow.forward(200)
20
21     #forward, left, forward
22     myArrow.left(120)
23     myArrow.forward(100)
24
25     #turn around and draw the third line
26     myArrow.backward(100)
27     myArrow.right(240)
28     myArrow.forward(100)
29
30  def leftArrow():
31      #setup the green, 20 arrow
32      myArrow.color("green")
33      myArrow.pensize(20)
34
35      #set its position and direction
36      myArrow.penup()
37      myArrow.setpos(0,0)
38      myArrow.setheading(180)
39      myArrow.pendown()
40
41      #forward for the main arrow body
42      myArrow.forward(200)
43
44      #forward, left, forward
45      myArrow.left(120)
46      myArrow.forward(100)
47
48      #turn around and draw the third line
49      myArrow.backward(100)
50      myArrow.right(240)
51      myArrow.forward(100)
52
53  while True:
54      #get a letter from the user
55      userEntry = input("Enter your direction, L or R: ")
56
57      #L = call the leftArrow function
58      if userEntry == "L":
59          leftArrow()
60          #wait 3 seconds, clear
61          time.sleep(3)
62          myArrow.clear()
63
64      #R = call the rightArrow function
65      if userEntry == "R":
66          rightArrow()
67          #wait 3 seconds, clear
68          time.sleep(3)
69          myArrow.clear()
```

Figure 2.23   Code.

Light Sensor to Detect When Vampire Zombies Are Awake

Vampire zombies only come out when it's dark. Let's create a light sensor that tells us when it's dark so that we know when to be on alert.

Algorithm

```
If light sensor detects dark then
      Display alarm
```

For the micro:bit, the alarm will be some text across the LEDs. For the Circuit Playground Express, we'll flash the lights.

SELECTION

In coding, we call the *if* statement in the above algorithm **Selection**. *If* statements are a big part of coding. We use them to check on things. I like to use the example:

```
If Lorraine = hungry then
      Feed Lorraine
```

In an *if* statement, you're checking a condition, in this case, in this case: is Lorraine hungry? If that condition is true, then you run the code inside the *if* statement: Feed Lorraine.

---

micro:bit

Build

The light sensor displays a number that represents how bright it is. For the micro:bit, what we need to do is figure out what number is dark. Time for some science!

Build Algorithm

```
On button A press
      Show light sensor
            number
```

We want to show the light sensor number while we move it around from a light spot to a dark spot.

Build Code

1. From the **Input** menu, drag out ***on button A pressed***.
2. From the **Basic** menu, drag out ***show number 0***.
3. From the **Input** menu, drag out ***light level*** and drop it onto the 0.

Once you've downloaded the code, put your micro:bit in position. This position needs to be someplace where it gets dark. I put mine in the window.

Important note: The first number you get will be 255. Your micro:bit could be in total darkness or blazing sunlight, and still the number will always be 255. The micro:bit needs some time to adjust to the light/darkness before it can give an accurate reading. Ignore the first reading.

Figure 3.1   Vampire zombie detector.

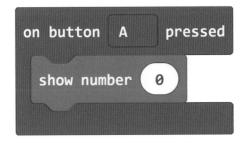

Figure 3.2   Show the number 0.

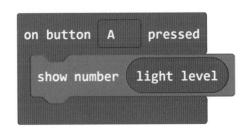

Figure 3.3   Show the light level number.

When it's bright out, press A, and write down the number. When it's dark out, press A again, and write down the number.

For my experiment, I got 250 for light and 10 for dark. So, I'm going to use anything less than 50 as my dark number.

Code

Now create the code to detect dark.

QUESTION: When do we want to check for darkness?
ANSWER: Forever!

The algorithm needs to go into a forever loop.

1. If it's not on your screen: from the **Basic** menu, drag out the *forever* block.
2. Let's build up our *if* statement, block by block.
   a. From the **Logic** menu, drag out *if true then* and place it inside *forever*.
   b. From the **Logic** menu, drag out *0 < 0* and place it on top of the word *true*.

Tip: I always forget which symbol is less than and which one is greater than. This year, my boss told me that less than is like the letter *L*! Thanks, Lynne!

   c. From the **Input** menu, drag out *light level* and drop it on top of the first 0.
   d. Change the second 0 to your dark number. Remember mine was 50.
   e. From the **Basic** menu, drag out *show string 'Hello!'*.
   f. Change *Hello* to *Dark*.

Debug

The main debugging you need to do with this mission is the light number. You need a number that will go off in the dark and not when it's a bit cloudy outside! Keep testing your micro:bit in different situations.

Expert Level

The word *Dark* scrolling past isn't very noticeable. Why not try to create an animation

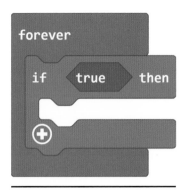

Figure 3.4 If true then block.

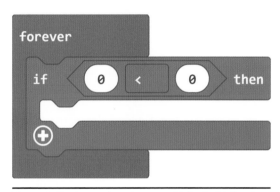

Figure 3.5 Is zero less than zero? I don't think so!

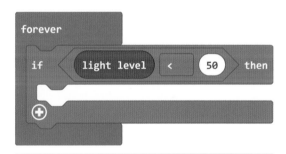

Figure 3.6 Is light level less than 50?

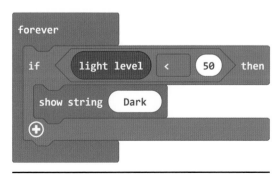

Figure 3.7 Checking for darkness.

of lights? I made the letters *S, O, S* flash one after the other 10 times. Can you create an animation using the **Loops** menu?

Circuit Playground Express

Build

The light sensor returns a number that represents how bright it is. For the Circuit Playground Express, we need to figure out what number is dark. Time for some science!

Build Algorithm

```
On button A Press
        Show light sensor number
```

Build Code

1. From the **Input** menu, drag out *on button A click*.
2. From the **Light** menu, drag out *graph 0* block and place it inside *on button A click*.
3. Press + on *graph 0*, and type 255 in place of 0.
4. From the **Input** menu, drag out *light level* and drop it on the 0.

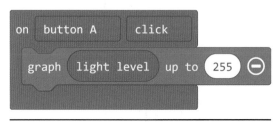

Figure 3.8 Graphing the light level.

Once you've downloaded the code, put your Circuit Playground Express in position. This position needs to be someplace where it gets dark naturally. I put mine in the window.

When it's daylight, press A and watch the lights. When it's dark out, press A again and watch the lights.

Each light on the Circuit Playground Express represents up to 25.5 units of light. For my experiment, I got eight lights on for light (204) and none for dark. So, I'm going to use anything less than 150 as my dark number. Watch out with this experiment! Daylight is very different from room light. Even a room with a light on is not as bright as daylight. You'll get very different numbers if you do your test in a room.

Code

QUESTION: When do we want to check for darkness?

ANSWER: Forever!

The algorithm needs to go into a forever loop.

1. If it's not on your screen: from the **Loops** menu, drag out the *forever* block.

2. Let's build up our Logic, block by block. We're going to check whether the light level is less than 150 to start us off.
   a. From the **Logic** menu, drag out *if true then* and place it inside *forever*.
   b. From the **Logic** menu, drag out *0 < 0* and place it on top of the word *true*.
   c. From the **Input** menu, drag out *light level* and drop it on top of the first 0.
   d. Change the second *0* to *150*.
   e. From the **Light** menu, drag out *set all pixels to red*.

Next, you need to experiment with your sensor. We set darkness as anything less than 150, but is that low enough? Put the sensor in a window at night and see if it turns red. If not, lower the number. Find a number that suits your light conditions.

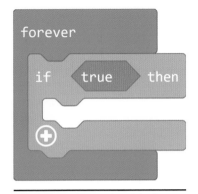

Figure 3.9 If true then block.

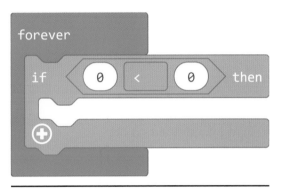

Figure 3.10 Is zero less than zero??

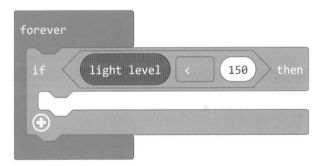

Figure 3.11 Checking the light level.

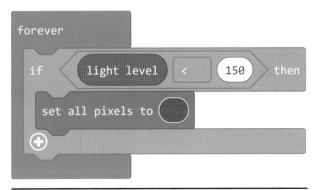

Figure 3.12 Show lights if dark.

Debug

I made the error of testing the light sensor in a room with a light, coding it but then placing it in sunlight. Getting the number that darkness occurs at is the tricky part in this mission.

A second bug is that when it's dark, the Circuit Playground Express will display red lights, but then it keeps displaying red, even when the sun comes up. We need to clear the lights when it's light.

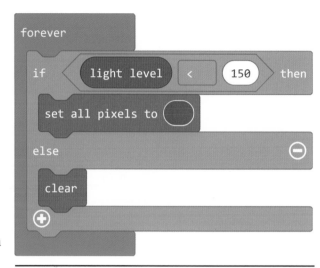

Figure 3.13   Debugged.

1. Press the + sign on the *if* block to add the *else* statement.
2. From the **Light** menu, drag out *clear* and place it inside *else*.

Now we have a vampire zombie alarm that will light up red when it's dark to tell us that the zombies are on their way. Then, in daylight, the lights will clear, and we can relax.

Expert Level

Red lights aren't very noticeable. Why not try to create an animation of lights? The Circuit Playground Express has a buzzer that makes noise. I made a police siren that flashes red and blue 10 times. Can you create an animation using the **Loops** menu? Find more ideas on the website (savetheworld.mcunderwood.org).

Light-Up Backpack

This backpack can be useful for lots of reasons—keeping you safe on your walk home from school, making your friends at school jealous, and so on. But we're going to use it to deter vampire zombies. Obviously!

Algorithm

Let's be a bit more fancy and use our backpack as an indicator! In this way, we don't have to leave equipment behind, and our friends can follow us. It's like a mobile Mission 2!

```
When I press A
       Indicate left
When I press B
       Indicate right
```

Figure 4.1   Super-cool backpack.

Build

The build for this mission is the same for the micro:bit and the Circuit Playground Express up until you attach wires to your device. You will need:

- A backpack
- Sewing needle and thread
- An RGB light strip of 30 lights
- Two male-to-male jumper wires (three for the Circuit Playground Express)
- Two crocodile clips (three for the Circuit Playground Express)

For the micro:bit, you'll also need:

- A 2 × AAA battery pack (one with a switch is better) and batteries
- Flat-head screwdriver
- Two terminal blocks, sometimes called *chocolate blocks* (nom nom)
  - You need two blocks attached together. They generally come in long strips. Pop off two blocks, keeping them attached together.
- A pair of scissors
- Wire strippers or a trusty knife

For the Circuit Playground Express you need all of the above, except make the battery pack 3 × AAA.

In the United Kingdom, I've found eBay to be a good cheap source for the strip of lights. You just need to search for WS2812B. You're looking for a 5-volt strip. I would go for the waterproof ones. In the United States, Adafruit sells many different types of NeoPixel strips that will work, which you can find on its website at www.adafruit.com.

Attach Wires

On one end of your light strip, there are several wires coming out of three pads. The pads are labeled and the wires are colored as follows:

Label	Color	What Is It?
+5V	Red	Power
DIN	Green	Data
GND	White	Ground

Your light strip might be colored and labeled differently. Draw the table above with the values for your light strip so that you don't lose track of what is what.

There are three wires: power, data, and ground. Think of a battery with two sides labeled + and − . The plus (+) is power, and the minus (−) is ground. Everything that uses electricity needs power and ground. Your light strip needs this extra third wire for data so that you can tell each light what color to be.

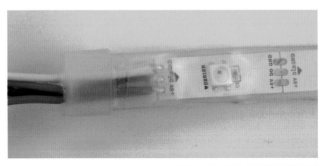

Figure 4.2   Power data and ground.

Other lights such as light bulbs go on and off; they have just two wires: power and ground. On my light strip, the three wires have a black connector attached.

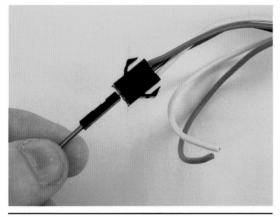

Figure 4.3 Blue jumper wire attaching to the green data wire.

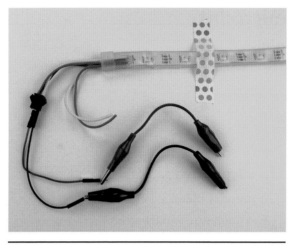

Figure 4.4 Crocodile clips attached.

1. Put one end of a male-to-male jumper wire in ground (white) and data (green).
2. Attach a crocodile clip to the other end of the jumper wires.

Those are the data and ground parts of the light strip done. We haven't attached anything to the 5V+ power wire yet. This step is different for each device.

Attach Everything to the Backpack

When you've added the power, how you attach the light strip depends on your backpack. My bag has a handy pouch in the back that I was able to fit everything into, including the battery pack(s). I cut a small hole in this pouch and wrapped the lights around. I sewed the light strip onto the edge of the bag with thread.

Note: It's really important that the metal parts of the crocodile clips and the jumper wires don't touch each other. Maybe tape them up before tucking them into your bag. This will also help secure the correct jumper wire to the crocodile clip.

If your bag is expensive, you might not want to use it. This light strip could be easily added to anything else, such as a sweater! This light strip is waterproof, but the bag and the electronic device aren't. I'd avoid going out in the rain wearing it.

micro:bit

Build

Let's attach the micro:bit to the light strip first.

Attach Data to Data

1. Find the data crocodile clip (mine is blue).
2. Attach it to pin 0 on the micro:bit.

Attach Ground to Ground

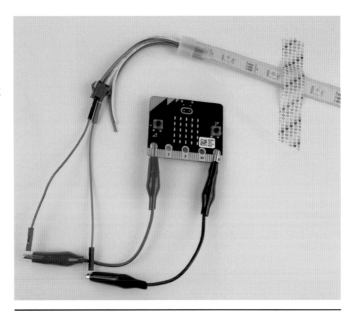

Figure 4.5   All connected.

1. Find the ground crocodile clip (mine is black).
2. Attach it to GND on the micro:bit.

Attach a Battery to the Light Strip

Now we need to power the light strip. The micro:bit can only reliably power up to eight lights at once from its own battery pack. It can power more, but you won't get the full colors, and your batteries will run out really quickly. My backpack has 30 lights, so I'm going to add an extra battery to the light strip. If yours has 10 lights or fewer, you don't need to do this step.

You can't add a bigger battery to the micro:bit; it won't take the power. You can add the battery

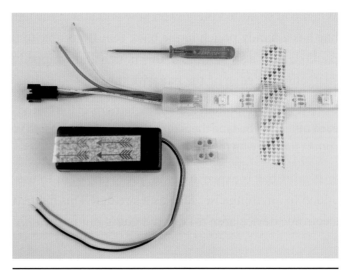

Figure 4.6   Data out end.

pack to either side of the light strip. For balance, I'm going to add it to the end opposite the micro:bit, the end where the wires are labeled:

- 5V+, power
- DO, data out
- GND, ground

Your light strip might be colored and labeled differently.

We're just looking for the wires coming out of 5V (red) and GND (white) to attach the battery to. In my strip, these are two separate wires not attached to a connector.

1. Strip back the red and white plastic to reveal the wire underneath on the light strip. You can get special wire strippers to do this, or ask an adult to help you do it with a knife.

TERMINAL BLOCK

Terminal blocks are great for connecting wires together. They are a plastic tunnel with two bolts on top, one at either end of the tunnel. When you screw down a bolt, it holds the wire in place and electrically connects it to the other bolt/wire. When you first get a terminal block, you may need to unscrew the bolt so that you can fit your wire in.

---

2. Place the red wire inside a hole in the terminal block. Screw down that block using a flat-head screwdriver. Keep tightening it until it won't go anymore. It might take a few attempts to get this right. Tug the wire gently to check that it's secure. Do the same with the white wire on the next block on the same side.

3. I'm going to use the same battery pack the micro:bit uses: a 2 × AAA battery pack. My battery pack has a switch on it, which is really useful for this mission. Make sure that there are no batteries inside the battery pack before you continue.

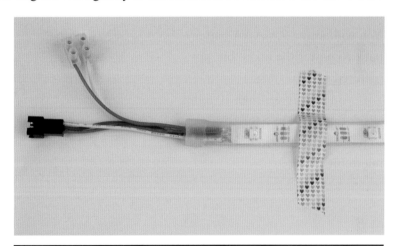

Figure 4.7 Both wires in the terminal block.

4. If it has one, cut off the connector at the end of the battery pack so that you have access to the two wires.

5. Strip back the red and black plastic covers of the battery wires.

6. Connect ground to ground.

   a. Screw the black wire of the battery pack into the opposite hole from the white wire of the light strip in the terminal block.

7. Connect power to power.

   a. Screw the red wire of the battery pack into the opposite hole from the red wire of the light strip in the terminal block.

We're attaching ground of the light strip to ground of the battery pack and power of the light strip to power of the battery. Repeat after me: Ground to ground. Power to power. That's our light strip all powered up!

That's it! Here's the entire strip with the micro:bit at one end and the battery pack at the other. *Note: You will need a 2 × AAA battery pack to power the micro:bit, not pictured here.*

If you have fewer than 10 lights, check out Mission 5 for how to power a light strip from the micro:bit.

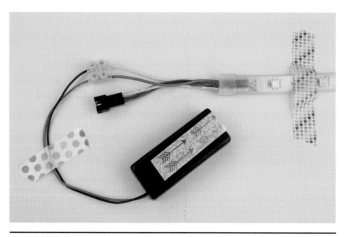

Figure 4.8   Battery attached.

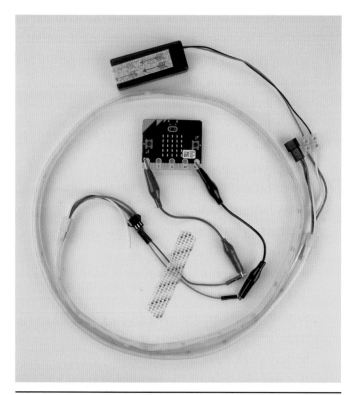

Figure 4.9   The entire strip.

Attach Everything to the Backpack

Here's the start and end of the light strip going into the bottom of the pouch.

Code

Let's re-create the code from Mission 2.

This is basically the code we need, but with an added extra of turning on lights on the light strip.

Build

Set Up the Light Strip

In the Raspberry Pi section of Mission 1, we talked about libraries. We're going to add the library to control the lights on the backpack.

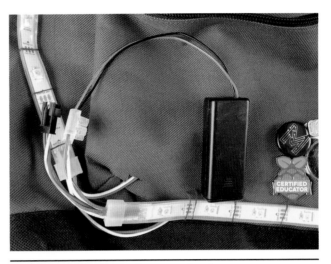

Figure 4.10 The light strip start and end.

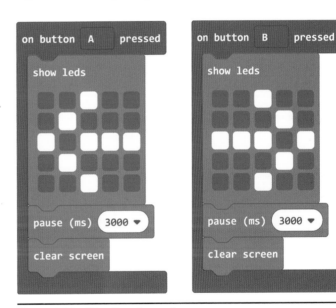

Figure 4.11 Direction sign code.

LIBRARIES IN MAKECODE

MakeCode doesn't have all the code it needs loaded all the time. Sometimes we need to tell MakeCode to go fetch the code we need.

MakeCode is similar. There are extensions that you can add to MakeCode. One of these extensions is the NeoPixel library.

1. From the side menu, scroll down and select **Advanced**.
2. Scroll down again and select **Extensions**.

    On this screen, you normally can see the NeoPixel extension. (If you can't see it, type *neopixel* in the search box.)

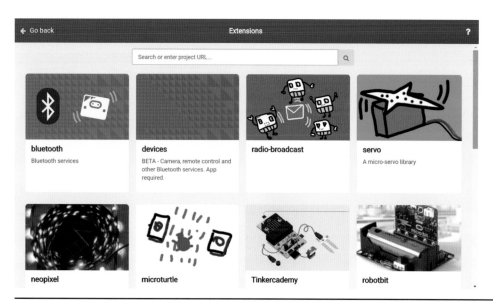

Figure 4.12   NeoPixel extension.

3. Select the NeoPixel extension. You will now have **NeoPixel** in your menu.
4. From the **NeoPixel** menu, drag out *set strip to NeoPixel at P0 with 24 leds as RGB (GRB format)* and place it inside *on start*. This creates a variable called *strip* and sets it to a strip of 24 Neopixels attached to pin 0 of the micro:bit.

Figure 4.13   Set up the NeoPixel strip.

VARIABLES

A *variable* is somewhere we can store information. A great example of a variable is your score in a game. The score can be increased and decreased, we can see the score, and if you're a big cheat, you can change the score altogether. The same goes for variables! Variables can also be different data types such as text and numbers. The ***strip*** variable is used to change the lights on our backpack. We can change its color.

---

5. Change 24 to the number of LEDs in your strip.

We can now change the colors of the LEDs using the ***strip*** variable in the **NeoPixel** menu. But we don't want to change the whole strip; we just want to change the left side or the right side.

Let's create a range of lights for the left indicator and the right indicator. When we color in the range, it will just color in those lights.

In my strip, left will be lights 0 to 14, and right will be lights 15 to 29.

1. In **Variables**, select ***Make a Variable.***
2. Type the variable name *Left.*
3. In **Variables**, select ***Make a Variable.***
4. Type the variable name *Right.*
5. From the **NeoPixel** menu, drag out ***Set range to strip range from 0 with 4 leds*** into ***on start.***
6. Change *range* to *Left.*
7. Change *4* to half the LEDs you have. I have 30 LEDs, so I'm going to change *4* to *15.*
8. From the **NeoPixel** menu, drag out ***Set range to strip range from 0 with 4 leds*** into ***on start.***
9. Change *range* to *Right.*
10. Change *0* to half the number of LEDs you have. I have 30 LEDs, so I'm going to change *0* to *15.*
11. Change *4* to half the number of LEDs you have. I have 30 LEDs, so I'm going to change *4* to *15.*

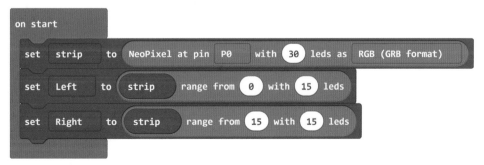

Figure 4.14 Set up Left and Right.

12. We can now change just the *Left* lights or just the *Right* lights.
    a. From the **NeoPixel** menu, drag out ***Strip show color red.***
    b. Change *Strip* to *Left* or *Right*.

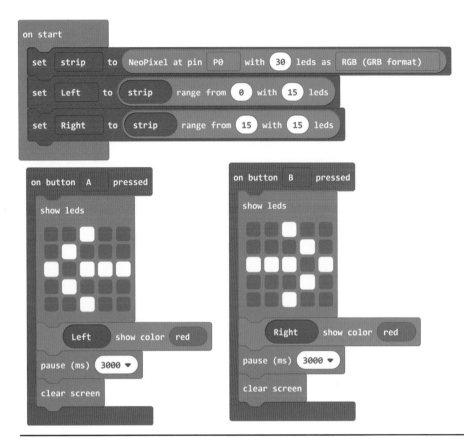

Figure 4.15   Indicators.

Of course, with a light-up backpack, you can create your own code and animations. Try out the block ***Strip show rainbow from 1 to 360*** under the **NeoPixel** menu. That's my favorite one!

Debug

We have a similar problem to what happened in Mission 2. The arrows clear after 3 seconds, but the LEDs do not.

From the **NeoPixel** menu, you need the blocks:

- ***Strip clear***
- ***Strip show***

Remember to change the word *strip* to the correct indicator: *Left* or *Right*.

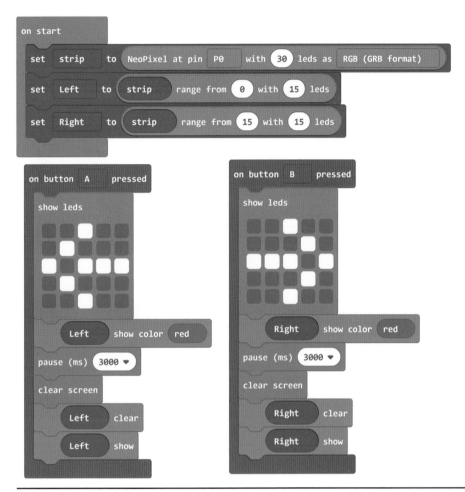

Figure 4.16 Clear lights.

In my bag setup, it's quite tricky to press the A and B buttons on the micro:bit when it's in the pouch. If I had a different bag, I'd try and add the micro:bit to my shoulder so that I could reach the buttons. Now that you know how to animate strips of lights, you could have any animation on there when the micro:bit starts. Or have a look at the next mission and see if you can copy the *tilt* code for your backpack.

Expert Level

Unlike the direction arrow, this strip of lights is an indicator. Indicators flash! Could you create a flashing light with the code above using a loop?

In Mission 12, you will learn how to use a second micro:bit to control another micro:bit. If you had a second micro:bit, you could create a remote for your backpack and turn it on from your hand!

Circuit Playground Express

Build

Let's attach the Circuit Playground Express to the light strip. Circuit Playground Express can power all the lights through the bigger 3 × AAA battery pack. We don't need to attach a second battery pack.

Attach Ground to Ground

Clip the other end of the GND (black) crocodile clip onto a GND pin on the Circuit Playground Express.

Attach Data to Data

Clip the other end of the DIN (blue) crocodile clip onto pin A1 of the Circuit Playground Express.

Attach Power to Power

1. Attach a male-to-male jumper wire (red) to the 5V+ connector (red).
2. Clip a crocodile clip onto this jumper wire.
3. Clip the other end of the 5V+ (red) crocodile clip onto pin VOUT of the Circuit Playground Express.

Figure 4.19 shows the whole strip with the Circuit Playground Express and the battery pack attached, all lit up!

If you find that your lights keep turning off when you move, try out the build in Mission 5. It's a bit more stable but requires chopping up crocodile clips.

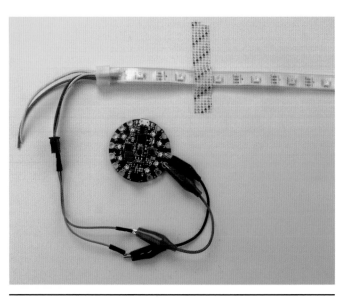

Figure 4.17   Both attached.

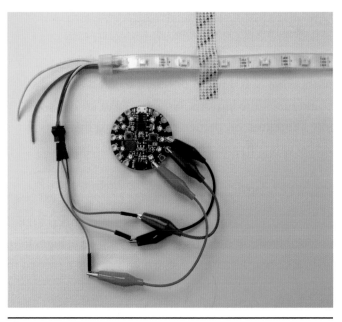

Figure 4.18   All done.

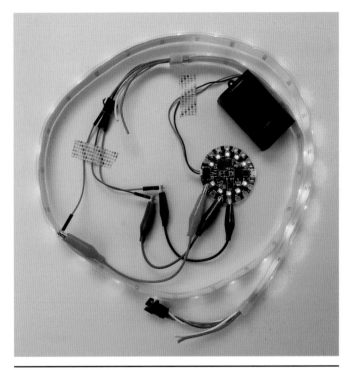

Figure 4.19 Entire strip.

Attach Everything to Your Backpack

Here's the Circuit Playground Express about to go into the pouch. Everything is turned on. Test that it still works before you tuck it all into the pouch.

Figure 4.20 Circuit Playground Express and battery about to go in the pouch.

Figure 4.21 Shiny lights.

Code

Let's take the code from Mission 2.

This is basically the code we need, but with an added extra of turning on lights on the light strip.

Set Up the Light Strip

1. From the **Loops** menu, drag out *on start*.
2. From the **Light** menu, select the submenu **NeoPixel**.
3. Drag out the block *Set strip to create strip on A1 with 30 pixels*, and place it inside **on start**. This creates a variable called *strip* and sets it to a strip of 30 NeoPixels attached to pin A1 of the Circuit Playground Express.

Figure 4.22   Direction sign code.

Figure 4.23   Set up the NeoPixel strip.

VARIABLES
Learn about *variables* from page 65 in this mission.

4. Change *30* to the number of LEDs in your strip.

We can now change the colors of the LEDs using the *strip* variable in the **NeoPixel** menu. But we don't want to change the whole strip; we just want to change the left side or the right side.

Let's create a range of lights for the left indicator and the right indicator.

1. In **Variables**, select *Make a variable*.
2. Type the variable name *Left*.
3. In **Variables**, select *Make a variable*.
4. Type the variable name *Right*.
5. From the **Variables** menu, drag out *Set right to 0* into **on start**.
6. Change *Right* to *Left*.
7. From the **Light > NeoPixels** menu, drag out *Strip range from 0 with 0 pixels*, and place it on the *0*.
8. Change the second *0* to half the number of LEDs you have. I have 30 LEDs, so I'm going to change *0* to *15*.
9. From the **Variables** menu, drag out *Set right to 0* into **on start**.
10. From the **Light > NeoPixels** menu, drag out *Strip range from 0 with 0 pixels*, and place it on the *0*.
11. Change both zeros to half the LEDs you have. I have 30 LEDs, so I'm going to change them both to *15*.

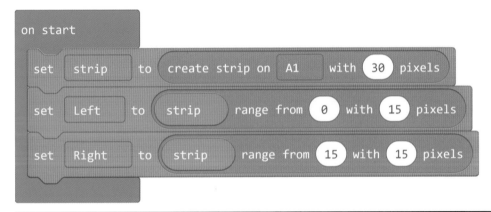

Figure 4.24　Set up left and right.

12. We can now change the *Left* lights and the *Right* lights.
    a. From **Light > NeoPixel**, drag out *Strip set all pixels to red*.
    b. Change *Strip* to *Left* or *Right*.

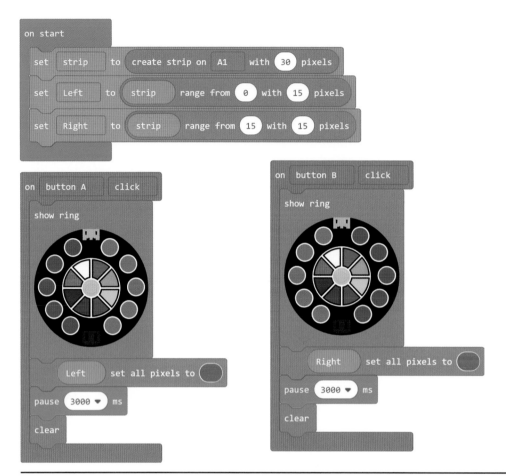

Figure 4.25 Indicators.

Debug

Lots of bugs here!

Bug 1

My lights don't seem that bright. You can increase the brightness of the light strip using the code *Strip set brightness 255.* The brightest setting is 255.

Figure 4.26 Shiny.

Bug 2

We have a similar problem to what happened in Mission 2. The arrows clear after 3 seconds, but the LEDs do not.

From the **NeoPixel** menu, you need the blocks

- ***Strip clear***
- ***Strip show***

Remember to change the word *Strip* to the correct indicator: *Left* or *Right*.

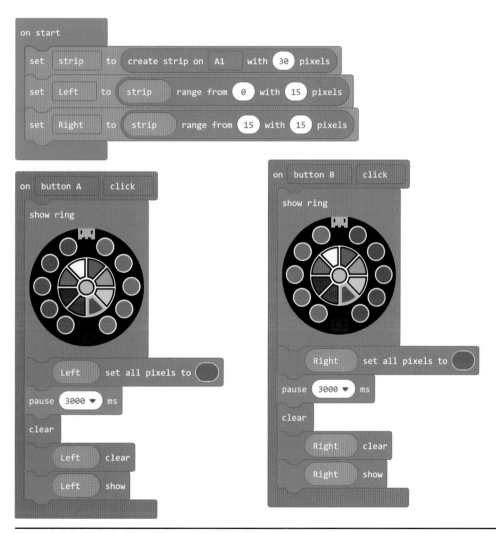

Figure 4.27 Clear lights.

Bug 3

In my bag setup, I can't reach the A and B buttons on the Circuit Playground Express when it's in the pouch. I could have tried to attach the Circuit Playground Express to a shoulder strap instead. I've changed the code so that an animation starts when the Circuit Playground Express starts. Have a look at Mission 5 to see if you can copy the *tilt* code for your backpack.

Expert Level

This light strip would look so much cooler if it flashed. Could you create a flashing light with the code above using a block from the **Loops** menu?

In Mission 12, you will learn how to use a second Circuit Playground Express to control another Circuit Playground Express. If you had a second Circuit Playground Express, you could create a remote for your backpack and turn it on from your hand!

Light-Up Attack Sword for Battling Vampire Zombies

This is the perfect combination for defending your home against vampire zombies: a light and a sword. With the micro:bit and the Circuit Playground Express, you can make a very cool sword. The question is, which one will win the battle?

Algorithm

We're not just going to light up a sword. We're going to use the tilt sensor inside each device and change the color of the light as we swing the sword.

```
On Left Tilt
      Go Green
On Right Tilt
      Go Blue
On button A press
      Make a cool animation
```

Figure 5.1 Choose your weapon wisely.

My cool animation is going to be turning on each light individually so that it looks like the sword is powering up. Let's create the algorithm for that as well.

```
On button A press
      Light up light 1
      Light up light 2
```

```
Light up light 3
  . . .
Light up light 8
```

This looks really inefficient! We can make this better by instead of doing eight statements of **Light up light**, we can use a loop. But we can't just use **Repeat 8 times**. We need to know what loop number we're on to turn on the correct light. A *for loop* is perfect for this.

A *FOR LOOP*

A *for loop* repeats a set number of times. It also gives you access to the current loop number through a variable.

With a *for loop*, our new algorithm looks like this:

```
For index counting from 0 to 7
    Set light number index to red
```

Index is a variable that stores which number the loop is currently on.

The first loop will be:

```
Set light number 0 to red.
```

The next ones will be:

```
Set light number 1 to red.
Set light number 2 to red.
Set light number 3 to red.
```

and so on until we reach 7.

Build

The build is the same for both devices up until you attach them. How cool is this?

Here's the equipment you'll need:

- A wooden sword
- A strip of lights
- Three crocodile clips
- Three terminal blocks still connected (optional)

Figure 5.2 Ready.

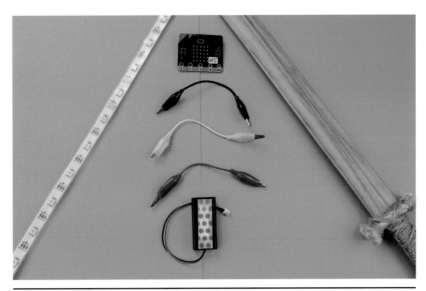

Figure 5.3 Equipment list; choose micro:bit or Circuit Playground Express.

Because the sword is going to move around a lot more than the backpack, I've got a different way to connect the devices to the light strip.

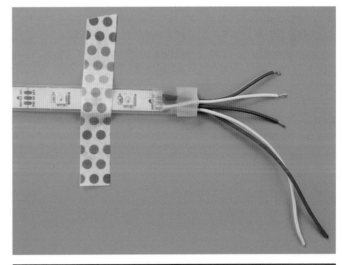

1. Find the **_data in_** end of the strip. This is the end labeled **DIN**.
2. Cut off any adaptors on the end.
3. Strip back the GND, DIN, and +5V wires as in Figure 5.4.
4. With the backpack, we clipped on the crocodile clips as in Figure 5.5.
5. With the sword, I think we need a more stable connection.

Figure 5.4 Use a wire strippers or (with some help) a knife.

6. Chop off one of the heads of each crocodile clip. This is where the sword becomes a more permanent project. If you don't want to ruin your crocodile clips, you can stick with the method above. Just tape up the wire and crocodile connection to try to make it more stable.

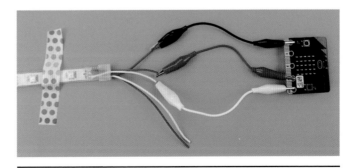

Figure 5.5 Backpack method.

7. Screw the stripped green, red, and white wires into each hole of the terminal block.

8. Screw the crocodile clips into the terminal blocks. The colors of the crocodile clips don't matter, but I like to match wires where I can. Your setup should now look like Figure 5.7.

On my light strip:

- Green is DIN, data in
- Red is +5V, power
- White is GND, ground

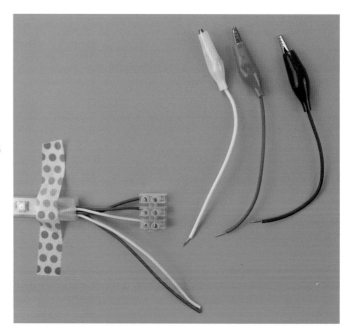

Figure 5.6   Off with their heads!

So I'll be attaching:

Light Strip	Wire Color	Crocodile Clip Color	micro:bit Pin	Circuit Playground Express Pin
DIN	Green	Green	0	A1
5V	Red	Blue	+3V	+3.3V
GND	White	White	GND	GND

Follow the steps for your device for the rest of the build.

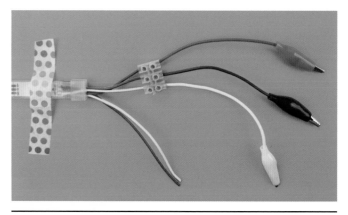

Figure 5.7   Croc clips ready to go.

micro:bit

Build

How you connect everything to your sword depends on the size of your sword. I've attached the micro:bit to one side and the light strip to the other.

I used double-sided tape to attach the battery onto the sword and normal tape to secure the battery in place. In Figure 5.8, the micro:bit is held in place using the battery and the crocodile clips. I'd recommend adding the cable ties as in Figure 5.11.

I've used double-sided sticky tape to stick down the light strip and the terminal block on the other side.

I've secured the micro:bit in place using zip ties.

- Tie one zip tie around the sword, just above the hilt.
- Loop a zip tie through pin 1 and the zip tie on the sword.
- Loop a zip tie through pin 2 and the zip tie on the sword.
- Cut off any excess zip tie plastic.

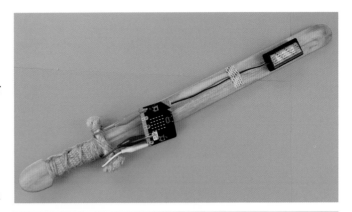

Figure 5.8 Side A.

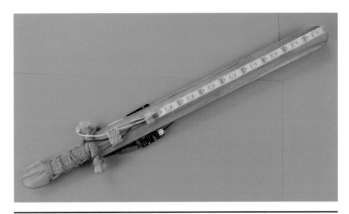

Figure 5.9 The light strip.

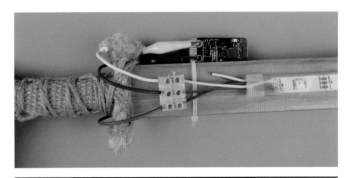

Figure 5.10 Terminal block up close.

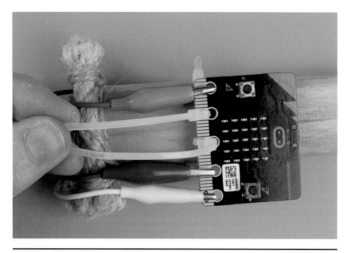

Figure 5.11 Pulling the zip ties tight.

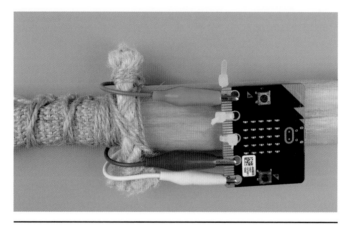

Figure 5.12 Tidy up the zip ties.

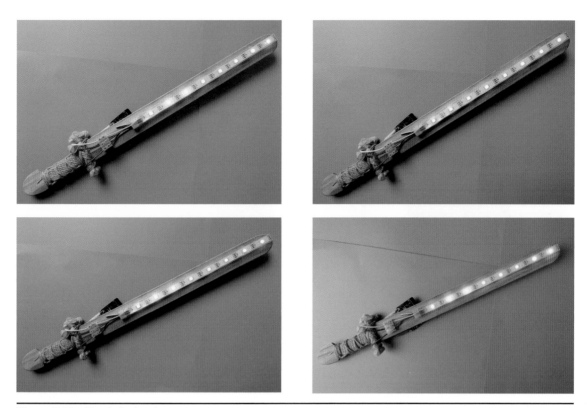

Figure 5.13 Ready for battle.

Code

1. Go to the Mission 4, "micro:bit" > "Build" > "Set Up the Light Strip," and follow steps 4 and 5 to set up the NeoPixel strip. If you haven't got the NeoPixel library, follow steps 1 to 3 first.
2. You should end up with the strip set to a strip of NeoPixel lights on pin 0.

Figure 5.14 Set up strip.

3. Change *24* to the number of LEDs you're putting on your sword. The micro:bit can only reliably power 8 LEDs from its own battery, so I'm sticking to 8.

Let's change the color of the entire strip when the micro:bit is tilted left.

1. From the **Input** menu, drag out *on shake*.
2. Select *shake*, and have a look at the movement options.

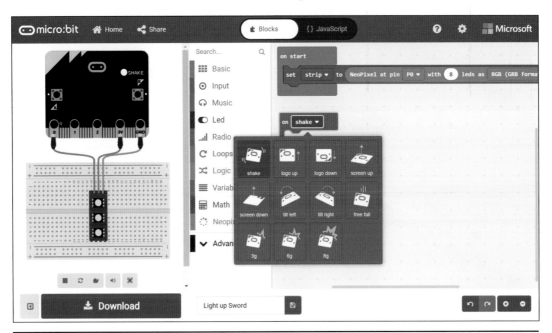

Figure 5.15   Different movements in MakeCode.

3. Select *tilt left*.
4. From the **NeoPixel** menu, drag out *strip show color red*, and place it inside *tilt left*.
5. Change *red* to *green*.

Let's change the color when we tilt right.

1. From the **Input** menu, drag out *on shake*.
2. Select *shake*, and change it to *tilt right*.
3. From the **NeoPixel** menu, drag out *strip show color red*, and place it inside *tilt right*.
4. Change *red* to *blue*.

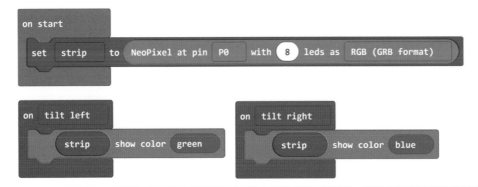

Figure 5.16   Tilt left and right.

Now let's make a cool animation! Remember our algorithm:

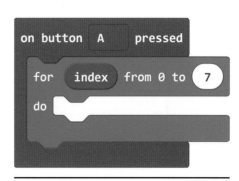

```
On A button pressed
        for index counting from 0 to 7

            set light number index to red
```

1. From the **Input** menu, drag out *on button A pressed*.

Figure 5.17 Cool animation loop.

2. From **Loops**, drag out *For index from 0 to 4 do*, and place it inside *on button A pressed*.

3. Change *4* to the number of LEDs you have on your sword minus 1.

COUNTING AT ZERO
In computing, we start counting at 0, so when you want to count to 10, the last number is 9. So if you want to count your 10 fingers, you would count 0, 1, 2, 3, 4, 5, 6, 7, 8, 9.

4. From the **NeoPixel** menu, select **More**, drag out the block *Strip set pixel color at 0 to red*, and place it inside the *for loop*.

5. From the **NeoPixel** menu, select *strip show*, and place it inside the *for loop*. The *block strip set…* prepares the NeoPixel strip; to apply the changes, you always need *strip show* after it.

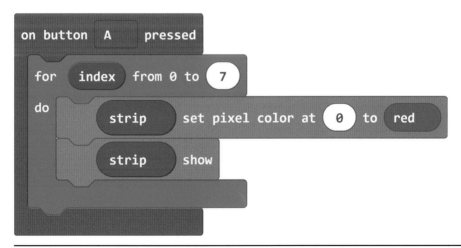

Figure 5.18 Show NeoPixels.

This code just sets light zero, eight times! We need to add the index variable into the loop.

6. From the **Variables** menu, drag out *index* and drop it on the *0* of *Strip set pixel color at 0 to red*.

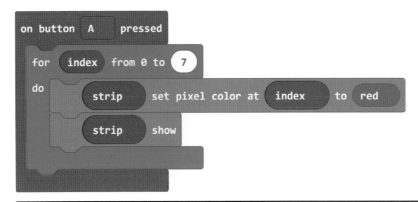

Figure 5.19 Variable added.

If my loop went from 0 to 4, this is the code that will run:

```
Strip set pixel color at 0 to red
Strip show
Strip set pixel color at 1 to red
Strip show
Strip set pixel color at 2 to red
Strip show
Strip set pixel color at 3 to red
Strip show
Strip set pixel color at 4 to red
Strip show
```

Here's all the code with the tilt and setup from earlier:

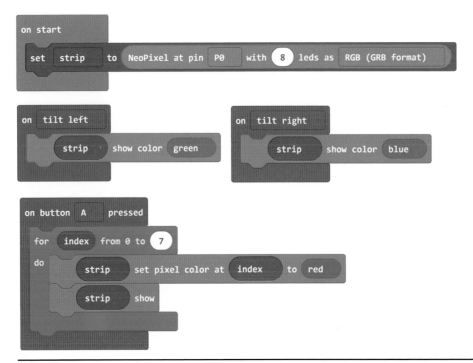

Figure 5.20 Final code.

7. Test the code in the simulator by clicking on the A button of the micro:bit on the screen.

Debug

Hopefully you'll have spotted in the simulator that the cool animation doesn't look very cool. The lights all turn red too fast. You can't see the power-up. Just put a pause in the *for loop*, a short one, like half a second.

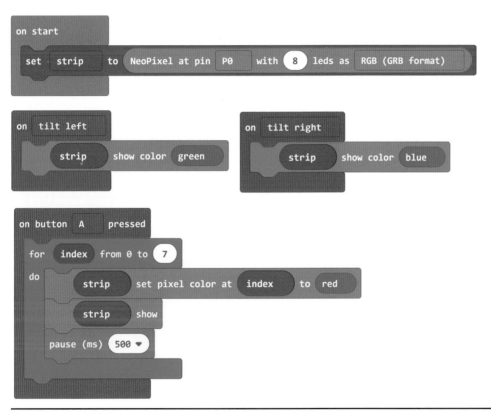

Figure 5.21   Add a pause.

Expert Level

Look at the other movements in MakeCode. Can you figure out which direction they are? Add some more colors to them! In Mission 9, you can create a lock for your sword.

Circuit Playground Express

Build

Here are the crocodile clips attached to the Circuit Playground Express on the sword.

How you connect everything to your sword depends on the size of your sword. I've attached the Circuit Playground Express to one side and the light strip to the other.

Figure 5.22   Attached.

I used double-sided tape to attach the battery onto the sword and normal tape to secure it in place. I haven't stuck the Circuit Playground Express on yet; it seems to be held in place by the battery and the crocodile clips, and then I used double-sided sticky tape again to stick down the light strip and the terminal block on the other side.

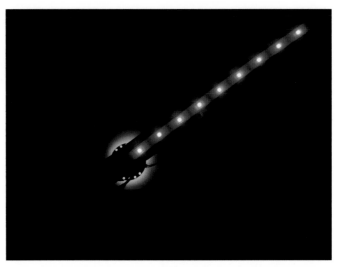

Code

Set Up the Light Strip

1. Go to Mission 4, "Circuit Playground Express" > "Build" > "Set Up the Light Strip," and follow steps 1 to 4.

Figure 5.23   Ready for battle.

2. Make sure that you've set the number of pixels to the number of lights on your sword. I've gone for 10.

Figure 5.24   Set up the strip.

Let's change the color of the entire strip when the Circuit Playground Express is tilted left.

1. From the **Input** menu, drag out *on shake*.
2. Select *shake*, and have a look at the movement options.

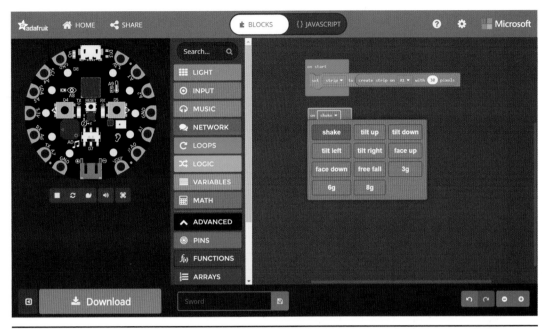

Figure 5.25   Looking for tilt.

3. Select *tilt left*.
4. From the **Light > NeoPixel** menu, drag out *Strip set all pixels to*, and place it inside *tilt left*.
5. Change *red* to *green*.

Let's change the color when we tilt right.

1. From the **Input** menu, drag out *on shake*.
2. Select *shake*, and change it to *tilt right*.
3. From the **Light > NeoPixel** menu, drag out *Strip set all pixels to*, and place it inside *tilt right*.
4. Change *red* to *blue*.

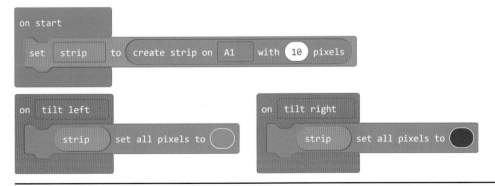

Figure 5.26   Tilt left and right.

Now let's make a cool animation! Remember our algorithm:

```
On A button pressed
        for index counting from 0 to 9
                set light number index to red
```

1. From the **Input** menu, drag out *on button A click*.
2. From the **Loops** menu, drag out *For index from 0 to 4 do*, and place it inside *on button A pressed.*
3. Change *4* to the number of LEDs you have on your sword minus 1. (See the box "Counting at Zero" on page 83 explaining why we're setting the LEDs to one less than the number of lights.)
4. From the **Light > NeoPixel** menu, drag out the block *Strip set pixel color at 0 to red*, and place it inside the *for loop*.

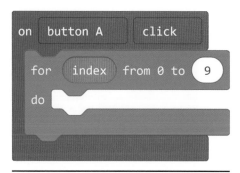

Figure 5.27 Cool animation loop.

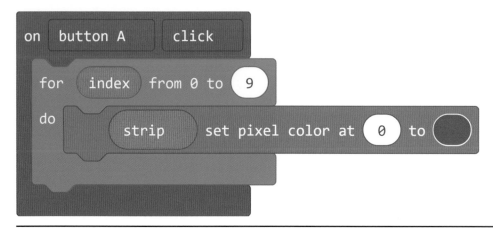

Figure 5.28 Show NeoPixels.

This code just sets light 0 ten times! We need to add the *index* variable into the loop.

5. From the **Variables** menu, drag out *index*, and drop it on the *0* of *Strip set pixel color at 0 to red.*

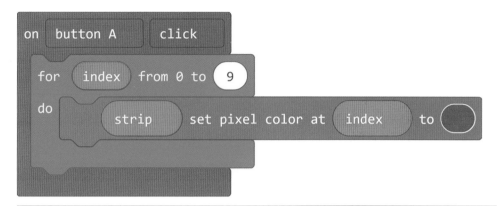

Figure 5.29   Variable added.

If my loop went from 0 to 4, this is the code that will run:

```
Strip set pixel color at 0 to red
Strip set pixel color at 1 to red
Strip set pixel color at 2 to red
Strip set pixel color at 3 to red
Strip set pixel color at 4 to red
```

Here's all the code with the tilt and setup from earlier:

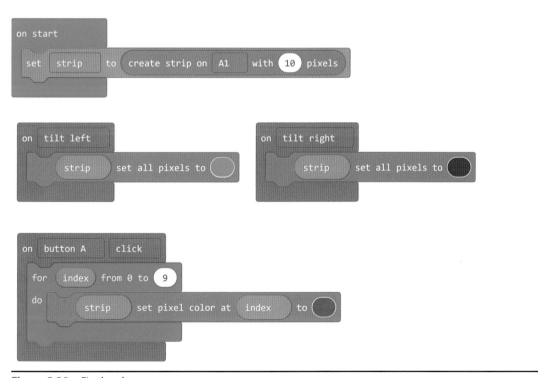

Figure 5.30   Final code.

Debug

When you press A, you'll see that the cool animation doesn't look very cool. The lights all turn red too fast, and you can't see the power-up effect. Just put a pause in the *for loop*, a short one, like half a second.

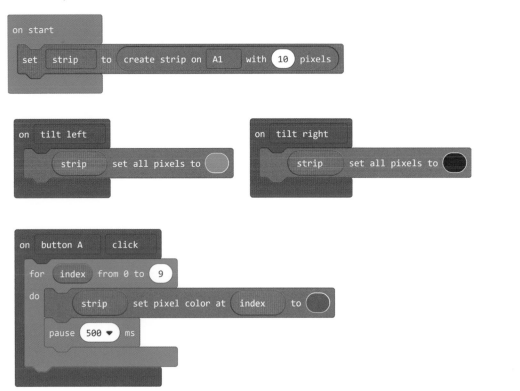

Figure 5.31   Add a pause.

Expert Level

Look at the other movements in MakeCode. Can you figure out which direction they are? Add some more colors to them!

Reaction Game to Test for Zombies

Everyone knows that zombies are really slow and can't react as quickly as humans. So a good test to know if someone is a zombie or not is to challenge him or her to a reaction game.

The game is simple: there are two buttons. The device counts down, and when it gets to 0, the first person to press his or her button wins, is human, and won't be hunted down. The following program uses the built-in buttons on the micro:bit and the Circuit Playground Express and the keyboard for the Raspberry Pi. The "Expert Level" adds cool arcade buttons that you can slam with your fist.

The game is simple, but the code is not! Let's write down the problem in a logical way.

Figure 6.1   I mean, how will we ever know who the zombie is?? They're identical twins!

Algorithm

```
Display 3, 2, 1 with a pause between each one
Clear Screen
If A is pressed:
            Show A
If B is pressed:
                Show B
```

But this algorithm won't work, because if button A is pressed then button B is pressed, button B will declare the winner.

We need the game to continue until someone presses a button. Whoever presses the button is the winner. Once a button is pressed, a winner is declared and the game ends. This is where a while loop comes in handy.

WHILE LOOP

A *while loop* is a loop that will continue until a certain event happens. We don't know how many times a while loop will loop; we just want it to stop when something happens—in this case, when a button, any button, is pressed.

When the button is pressed, we have our winner.

Full Algorithm

```
Display 3, 2, 1 with a pause between each one.
Clear Screen.
While button A is not pressed AND button B is not pressed
     Wait.
If button A was pressed
     A is the winner, show
        A/green.
If button B was pressed
     B is the winner, show
        B/blue.
```

Let's get coding!

micro:bit

Build

For this mission, make sure that the suspected zombies can reach their buttons and see the screen.

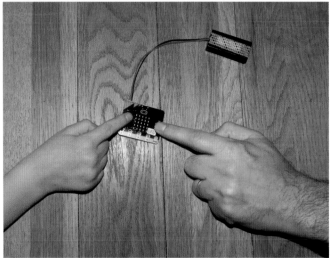

Figure 6.2 Testing Dad.

Code

Let's do the countdown first. Put all these blocks under a **forever** loop, and test it in the simulator. It should be counting down from 3 to 1 and then go back to 3.

Let's play until someone presses a button. Now let's grab our **while true** loop from the **Loops** menu and place it under **clear screen**.

1. From the **Logic** menu, pull out the block **and**, and place it on the *true* of the *while* block.
2. From the **Logic** menu, pull out the block **not**, and place it on the first diamond of the *and* block.
3. Do the same on the second diamond so that it looks like Figure 6.6.
4. From the **Input** menu, scroll down and select **button A is pressed**, drop it on the first **not** block's diamond. This can be tricky. Make sure that it's going on the diamond. Sometimes I accidentally replace the whole **not** block.

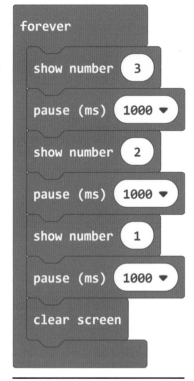

Figure 6.3 Countdown.

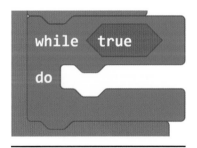

Figure 6.4 While.

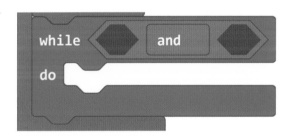

Figure 6.5 While and.

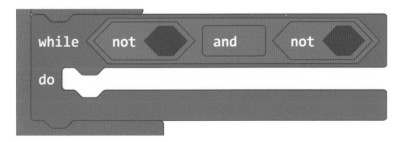

Figure 6.6 While and not.

5. From the **Input** menu, scroll down and select *button A is pressed*, drop it on the second *not* block's diamond. Change *A* to *B*.

Figure 6.7   While and not buttons.

6. So we're going to just wait until someone presses a button. Let's put a really short *pause* in here. We don't want to do anything while we're waiting, but MakeCode needs a block inside this loop. A really short pause won't affect the program.

Figure 6.8   While pause.

Who Won?

The next block after this *while* loop only gets run when a button is pressed. So let's check which button was pressed.

1. From the **Logic** menu, drag out *if true then.*

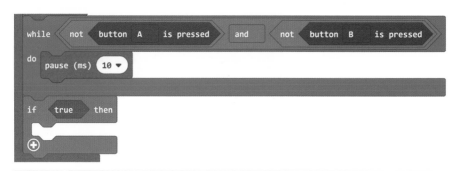

Figure 6.9   If.

2. From the **Input** menu, drag out ***button A is pressed***, and drop it on top of *true.*

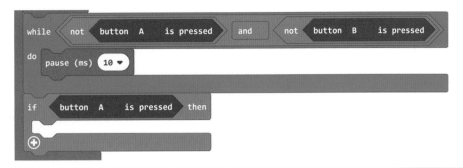

3. From the **Basic** menu, drag out ***show string "Hello!"***, and place it inside this **if** statement.
4. Click on *Hello*, and change it to *A.*

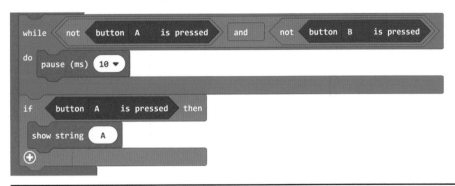

5. Do the same for the B button, replacing *Hello* with *B.*

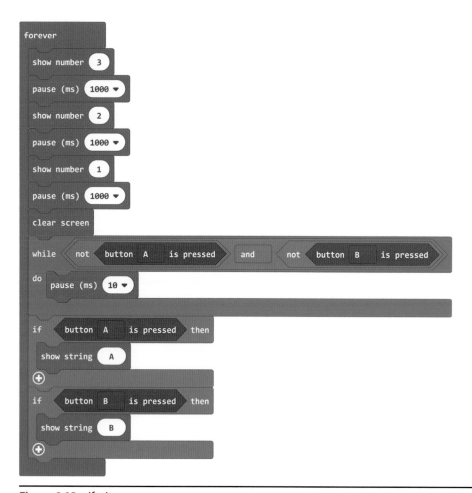

Figure 6.12   If winners.

That's it! Try out the game on the simulator before you download it to the micro:bit.

Before you download, give your project a good name at the bottom of the screen.

If you do the "Expert Level," you'll need to find this code again.

Debug

Something's not quite right—when I win, I want to cheer! But this game goes too fast. I could almost miss who won! How could you fix this?

Put a ***pause*** block after you display A and B. My pause is 3,000 milliseconds, which is 3 seconds. How long do you need to cheer?

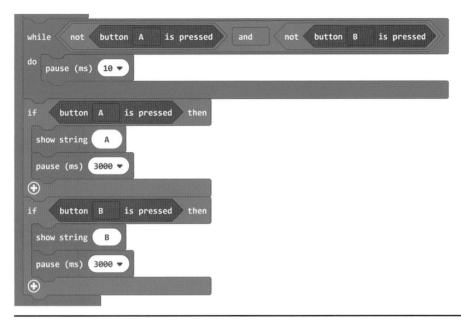

Figure 6.13 Pause.

Expert Level

Here's a cool build and code extension. Add some arcade buttons so that your human subjects can really smash the button! Find the full build and code steps for this on the website (savetheworld.mcunderwood.org).

Circuit Playground Express

Build

For this mission, make sure the suspected zombies can reach their buttons and see the lights. Staring down your opponent can also help you win.

Code

Use the lights around the Circuit Playground Express to indicate who has won and as a countdown clock. The countdown clock will be in red, player A will be green, and player B will be blue. I think you can follow

Figure 6.14 Testing Dad.

the steps in the micro:bit section to create this code. Everything is under the same menu in Circuit Playground Express. Here's your final piece of code:

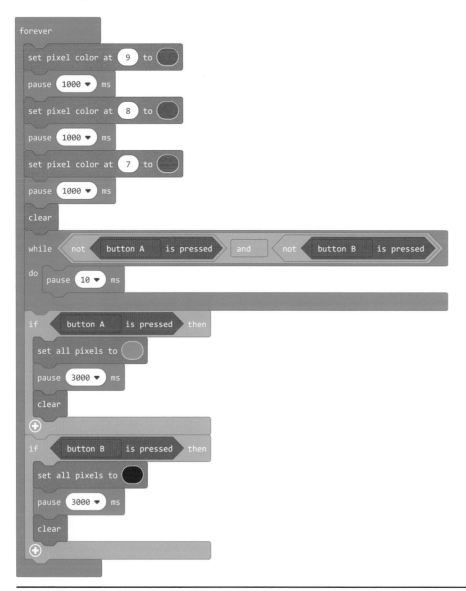

Figure 6.15   Final code.

Expert Level

To create a really cool reaction game, add some arcade buttons. Find the full build and code steps on the website (savetheworld.mcunderwood.org).

Raspberry Pi

We can use the keyboard for the reaction test. Let's use keys *A* and *L* on the opposite sides of the keyboard.

Algorithm

With some extra room, we're going to change the algorithm a bit for the Raspberry Pi.

```
Display "Ready", "Steady", "Go"
Get key press from keyboard
IF key was A
     Display "A wins!"
IF key was L
          Display "L wins!"
```

Build

Set up the Raspberry Pi with a screen, keyboard, and mouse attached. Make sure that the suspected zombies can reach their buttons and see the screen.

Code

Let's start!

1. Select *Programming > Python 3*.
2. Select *File > New*.
3. Start with a countdown (you'll need the time library. Learn what a library is from Mission 1, page 33.
4. Get the entry from the keyboard using the input function. I'm storing it in the variable ***userEntry***.
5. Now let's check who has won.

The players will have to press ENTER on the keyboard after they've tried to type a letter. That's it! Or is it?

```
1  import time
2
3  print ("Ready")
4  time.sleep(1)
5  print ("Steady")
6  time.sleep(1)
7  print ("Go!")
8  userEntry = input()
```

Figure 6.16 Countdown.

```
1   import time
2
3   print ("Ready")
4   time.sleep(1)
5   print ("Steady")
6   time.sleep(1)
7   print ("Go!")
8   userEntry = input()
9
10  if userEntry == "a":
11      print ("A wins!")
12  if userEntry == "l":
13      print ("L wins!")
```

Figure 6.17 Who won?

Debug

If the users both enter a letter, it will look like *AL* or *LA*, depending on who was the fastest. But the code is only looking for *A* or *L*. Nothing will happen if you enter *AL* or *LA*.

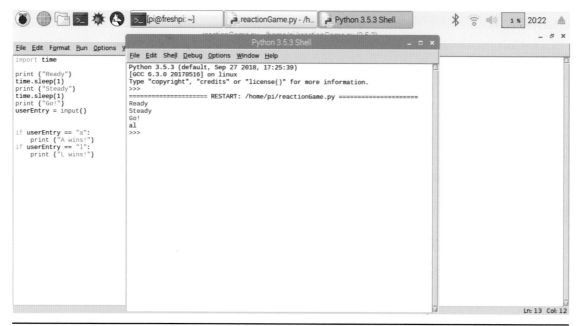

Figure 6.18   No one wins.

What we need the code to do is to look at the first letter of what was entered. We can use square brackets to do that. On my keyboard they're next to *P*. *userEntry[0]* gets the first letter of the variable *userEntry.*

When we run this code, a winner will be declared.

```
1  import time
2
3  print ("Ready")
4  time.sleep(1)
5  print ("Steady")
6  time.sleep(1)
7  print ("Go!")
8  userEntry = input()
9
10 #Check the first letter
11 if userEntry[0] == "a":
12     print ("A wins!")
13 if userEntry[0] == "l":
14     print ("L wins!")
```

Figure 6.19   Check the first letter.

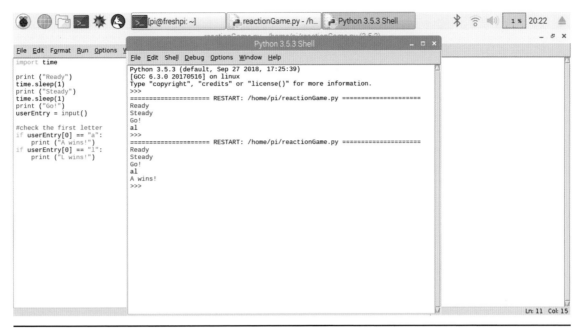

Figure 6.20   Who wins?

Also, let's put it all in a **while** loop so that we can keep playing forever! Remember to indent everything after *while True:*.

```python
1  import time
2
3  while True:
4      print ("Ready")
5      time.sleep(1)
6      print ("Steady")
7      time.sleep(1)
8      print ("Go!")
9      userEntry = input()
10
11     #Check the first letter
12     if userEntry[0] == "a":
13         print ("A wins!")
14     if userEntry[0] == "l":
15         print ("L wins!")
```

Figure 6.21   Code.

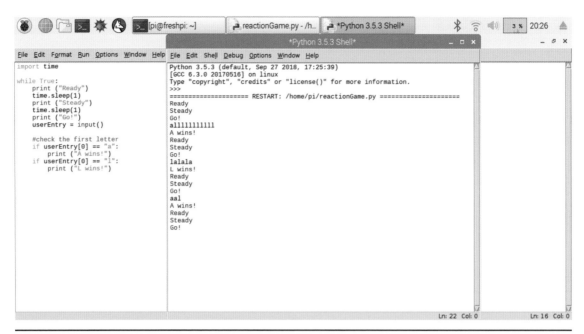

Figure 6.22   Program running.

Expert Level

You can easily add buttons to the Raspberry Pi. Check out the website for instructions on how to create an arcade-style reaction game (savetheworld .mcunderwood.org).

Figure 6.23   My suspects got really nervous.

PART TWO

Defend Your Home

Zombies have been discovered, kept away, and defeated. Now let's focus on your home and keeping it and your stuff safe. This section contains more builds, more complicated code, more missions, and more fun!

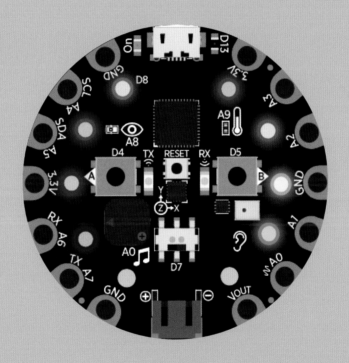

Name Badge to Identify Valid Family Members

Here's a nice, easy mission to start this section with: a name badge. This is to stop any spies from dressing up as your Dad and trying to enter your home. No name badge, no entry.

We'll use the micro:bit to scroll a name across the LEDs, and we'll use the Circuit Playground Express's colored lights to make a unique secret pattern.

Before we start coding, we're going to create our algorithm. Remember, an algorithm is a sequence of steps or rules for solving a problem. It's not code; it's like a plan for what we're going to do next. The algorithm will be similar for all our devices.

Algorithm

```
On start
        Display name/light pattern
```

micro:bit

Build

To attach the micro:bit as a badge, take out the USB cable, and plug in the battery. Pop the battery pack in a shirt pocket, and dangle the device out the front. The micro:bit sits quite nicely in a shirt pocket. If you don't have a shirt pocket, you could pop the battery pack into a T-shirt collar.

Alternatively, put some string through the holes in the micro:bit, and tie it carefully

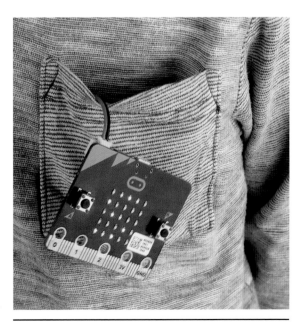

Figure 7.1   S for Sam!

105

around your neck as a necklace. Don't wear the micro:bit necklace to bed, and don't use metallic string.

Code

In this code, we're going to show the string "Lorraine" when the micro:bit starts. Replace Lorraine with your name, your family member's name, or your pet's name you want to identify.

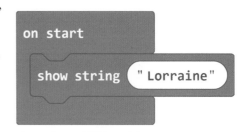

Figure 7.2   Show "Lorraine."

1. Test your code in the simulator.
2. If it's working and you're happy, download it to your computer and then transfer it to your micro:bit.

Debug

The problem with this code is that it will only run once when you start the device. To run it again, you will have to press the Reset button or take the battery out and put it back in again. Let's change it so that the name appears every time we press the A button.

Figure 7.3   Debugged.

Expert Mode

Add a secret message to the other button. Here's the algorithm. You create the code.

```
On button A pressed
        Display "Lorraine"
On button B pressed
        Display secret message
```

Circuit Playground Express
Build

To attach the Circuit Playground Express as a badge, take out the USB cable, and plug in the battery. Put the battery pack in a shirt pocket, and dangle the device out the front, just like the micro:bit picture earlier (Figure 7.1). Or pop the battery pack in a T-shirt neck, and let it dangle out the front like this!

Figure 7.4   Favorite colors!

Code

The Circuit Playground Express has a ring of lights around the edge. These lights can be lots of different colors. For our name badge, we will flash all the lights in three different colors as a secret code. Only the right colors in the right order will let you in the house. What colors will you choose?

Debug

There are two problems here! If you tested your code earlier in the simulator, you will have spotted the second problem. The first problem with this code is that it will only run once when you start the device. To run it again, you will have to press the Reset button or take the battery out and put it back in again. Let's change it so that the name appears every time we press the A button.

There's one final bug in this code. Did you spot it? Try pressing A in the simulator. What happened? Why? In my simulator, the Circuit Playground Express only displayed orange. Why do you think this is? Why didn't it show green or white?

Computers are really fast. The Circuit Playground Express *did* show green, white, and then orange. It just did it too fast for you to see. In order for us to see these colors, we need a pause between each one.

Figure 7.7 shows the new code to display green for a second, white for a second, and then orange.

Test it again in the simulator. If you're happy, download the code to the Circuit Playground Express. Give it to your trusted family members, and check their IDs from now on.

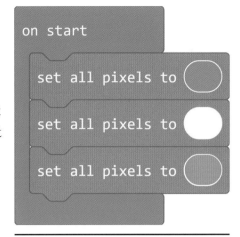

Figure 7.5   What are your favorite colors?

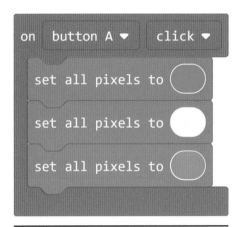

Figure 7.6   Debugged.

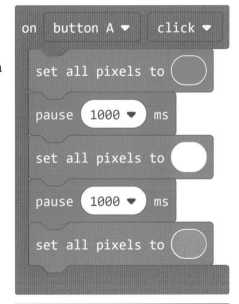

Figure 7.7   Final code.

Expert Level

Let's add a tune to the other button. Maybe it's their favorite song, or maybe you can make up a tune for them?

```
On button A pressed
     Display secret color codes.
On button B pressed
     Play secret tune.
```

See my secret tune on the website (savetheworld.mcunderwood.org).

Door Sensor to Tell When Your Room Is Under Attack

Little brothers, big sisters, none of them belong in your room—unless, of course, you share it with them. If you want to keep your home safe and siblings out your room, then I'd recommend the cookie jar lock for your belongings. This door alarm can work on any door, so you could add it to a closet or wardrobe instead.

The door sensor works on electrical signals. When two wires are touching, they let electricity flow from one to the other. If that connection is broken, then we can sense that electrically.

Back in Mission 1, you created a touch sensor to test whether a family member is a zombie. This worked because humans conduct electricity. Now we can't stay in our room all day waiting for a sibling to attack. So we're going to leave a conductive circuit behind.

One item that conducts electricity and that you can find in your home is tinfoil.

Let's look at the algorithm first and then get building.

Algorithm

```
If door is opened/circuit is broken
        Sound the alarm
```

Build

The build for this mission is the same until you connect your individual device.

Let's add a small speaker using crocodile clips. You will need:

- A speaker with a headphone jack and its own power source
- Four crocodile clips
- Tinfoil
- Tape

Add the Speaker

Mission 4 talks about how to use crocodile clips.

1. Add a crocodile clip to the tip of the headphone jack (yellow crocodile clip).
2. Clip the other end of this crocodile clip onto the GND pin.
3. Add a crocodile clip to the base of the headphone jack (green crocodile clip).
4. Clip the other end of this crocodile clip onto
   a. Pin 0 on the micro:bit.
   b. Pin A0 on the Circuit Playground Express.

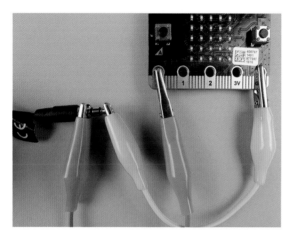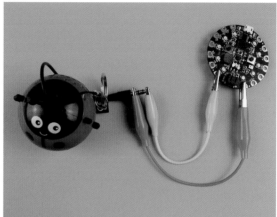

Figure 8.1   Adding a speaker to a project.

This will work connected the other way around as well. Use this test code to check that it works:

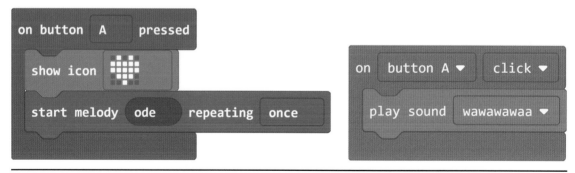

Figure 8.2   Testing the speaker on the micro:bit (left) and the Circuit Playground Express (right).

Add the Door Pads to the Door

1. Cut out two pieces of tinfoil. I started with two pieces each 6 × 8 inches.
2. Fold the pieces into thin rectangular strips. My final pieces were 4 × 2 inches.
   I wouldn't add them to the door frame until you've coded them first! It's much easier to test this project on your desk rather than on the floor.
3. Add one piece to the door frame using the tape.
4. Add the second piece to the door. Make sure that when you close the door, the two pieces touch.

Add the Door Pads to the Device

1. Add a crocodile clip to the tinfoil on the door frame. Attach the other end of the crocodile clip to pin 1 of the micro:bit or pin A1 of the Circuit Playground Express.
2. Add a crocodile clip to the other piece of tinfoil on the door. Attach the other end to GND of the micro:bit or the Circuit Playground Express. It doesn't matter which piece of tinfoil goes to GND or pin 1/A1, the door or the frame.

Here are all the crocodile clips and battery attached:

Figure 8.3 All the wires!

micro:bit

Code

Instead of pressing the pin, like we did in Mission 1 to test for zombies, we're looking for a block that detects the pin being released.

1. Under the **Input** menu, click on the submenu **More…** for the block *on pin P0 released*.
2. We're going to change P0 to P1 because we want to play music out of P0.

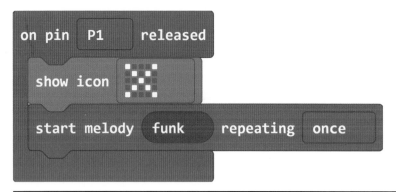

Figure 8.4   I like funk melody. What will you choose?

When testing the code, the micro:bit's pins work better when it's plugged into a battery pack instead of the USB on your computer.

This melody repeats just once—how many times do you want it to play? How long will it take you to get from the other side of the house to your room? I counted six melodies. Let's put it in a *repeat* loop instead.

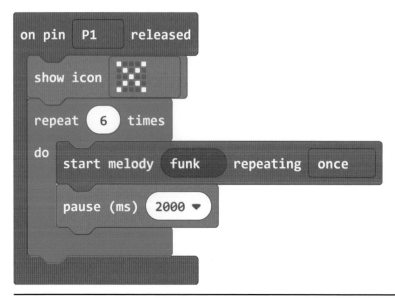

Figure 8.5   This creates a long alarm.

REPEAT LOOP
A *repeat* loop is a nice, simple loop. It just repeats the number of times you want it to.

When I first tested this, the melody sounded like it played only once. The loop was repeating the melody, but it was happening so fast that it sounded like one melody. I want one melody played six times, one after the other.

I put a pause after the melody to let it stop before it plays it again. The pause is 2,000 milliseconds, or 2 seconds, long. It lets the first melody finish before the next one begins.

Play around with your melody and pauses in the simulator to get the best alarm for your room.

Debug

They've broken into your room, you've dealt with them swiftly, but we've found a bug: the screen still shows an X on it. Let's reset the screen.

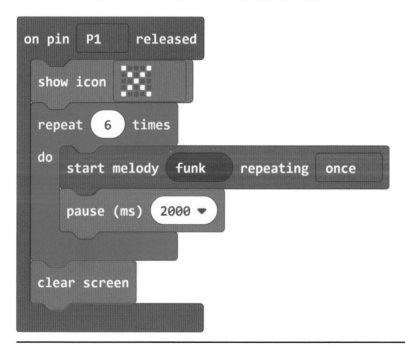

Figure 8.6 Reset.

When you debug this mission, it will mostly be debugging the build. Getting two pieces of tinfoil to touch is harder than it sounds. Putting them in the right place and making sure that the croc clips don't fall off—this is a delicate operation!

To help me debug, I kept the tune *on button A pressed*.

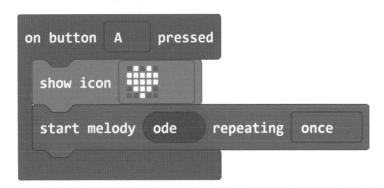

Figure 8.7   Debugging the speaker.

That way, I could test whether the speakers are set up correctly, separate from the door setup.

Expert Level

I have crafty brothers. I know they'll find a way around my alarm. Maybe they'll unplug the speaker. Check out Mission 12 on how to add a remote speaker. Also, why not use a variable to record every time the door is open. Take a look at the following algorithm. Can you code it?

```
On button A pressed
      Show doorOpened
On Pin P1 released
      Show an X
      Play a melody
      Change doorOpened by 1
      Show doorOpened
```

This way, every time you open your door, the micro:bit shows you the number of times the door has been opened. If it was more than the last time you opened the door, crafty siblings are at work!

When we have a mission like this that could stay set up for days, you'll find that your batteries will need replacing. I found this item from Monk Makes really helpful for long-term projects:

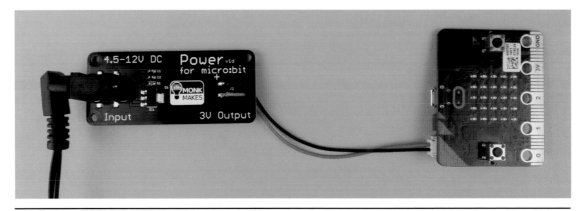

Figure 8.8   Power for micro:bit.

It lets you plug the micro:bit into a wall socket. You won't need batteries. Find out more information at https://www.monkmakes.com/mb_power/.

Circuit Playground Express

Code

Instead of pressing the pin down, like we did in Mission 1 to test for zombies, we're looking for a block that detects the pin going up. This is the same block; you just change *down* to *up*!

We also need to pull the pin up when the Circuit Playground Express starts.

The siren repeats just once. How many times do you want it to play? How long will it take you to get from one side of your house to your room? I counted four sirens. Let's put the ***play sound siren until done*** block inside a ***repeat*** block.

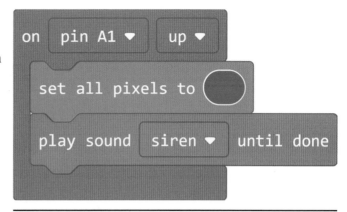

Figure 8.9   The siren is a great alarm sound.

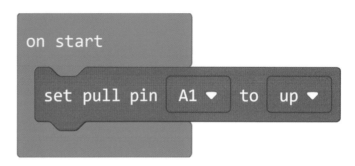

Figure 8.10   Set up the pin.

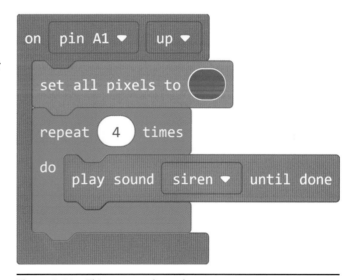

Figure 8.11   This creates a long alarm.

Debug

When I caught someone in my room and dealt with that person, I noticed that the Circuit Playground Express was still showing red. Let's clear the screen, ready for the next intruder.

Other problems with this mission are to do with the build. Getting two pieces of tinfoil to touch when the door closes is harder than it sounds. It's quite a tricky setup to get right. To help me debug, I kept the tune on a button click. That way, I can test whether the speakers are set up correctly, separate from the door setup.

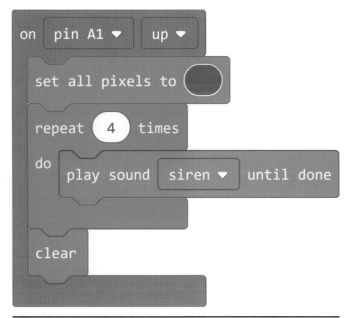

Figure 8.12   Reset.

To set up this mission, I found that if you open the door, press Reset on the Circuit Playground Express, and then close the door, this arms it, and it will recognize the door being opened. If the door is already closed when you start up the Circuit Playground Express, it won't recognize that the pin has gone up, because it didn't spot the pin going down.

Figure 8.13   Debugging the speaker.

Expert Level

There are crafty intruders out there, and they'll find a way around my alarm. Mission 12 shows you how to add a remote speaker so that the intruders can't remove your speaker and get in. Also, why not use a variable to record every time the door is opened. Take a look at the following algorithm. Can you code it?

```
On button A pressed
      Show doorOpened in LEDs
On Pin A1 released
      Show red lights
      Play a melody
      Change doorOpened by 1
      Show doorOpened in a graph
```

This way, every time you open your door, it shows you the number of times it has been opened. If it was more than the last time you opened it, there was an intruder!

Raspberry Pi

Build

The Raspberry Pi can play sound through the HDMI cable attached to the monitor or through a speaker you plug into it. For this mission, I'm going to attach a speaker and run the Raspberry Pi without a monitor.

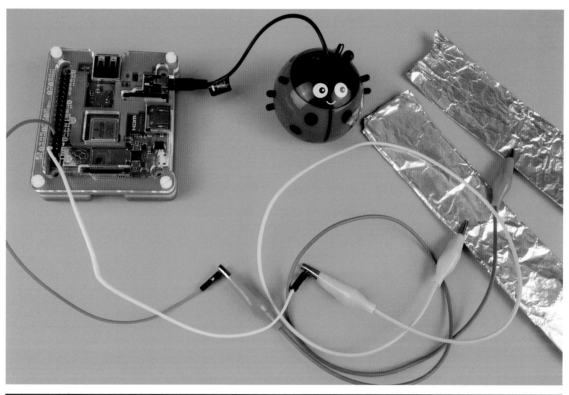

Figure 8.14 Door sensor with funky speaker.

You will need:

- Two male-to-female jumper wires
- Two crocodile clips
- Tinfoil
- A cool speaker

Use Figure 8.14 to guide you.

Connect a Ground Pin on the Raspberry Pi to a Piece of Tinfoil

1. Attach the female end of one jumper wire to a ground pin on the Raspberry Pi.
2. Attach a crocodile clip to the other (male) end of the jumper wire.
3. Attach the other end of the crocodile clip to a piece of tinfoil.

Do the Same for Another Piece of Tinfoil and GPIO2

1. Attach the female end of a jumper wire to GPIO2 on the Raspberry Pi.
2. Attach a crocodile clip to the other (male) end of the GPIO2 jumper wire.
3. Connect the other end of the crocodile clip to a piece of tinfoil.

Keep everything on your desk so that you can test it as you code it.

Code

Let's set up the code using a monitor first.

Set Up the Speaker

Download a siren noise to use as the alarm.

1. On the Raspberry Pi, open up the terminal (remember what the terminal is from Mission 1).
2. Type *wget 'http://soundbible.com/grab.php?id=1577&type=wav' -O siren.wav.* This should download a siren audio file called *siren.wav* to your Raspberry Pi. It *should*! But you might type it wrong, which I do all the time. This code is downloading a wav file from a website. If you find it easier, you could:
   a. Open up a browser on the Raspberry Pi.
   b. Type this URL: http://soundbible.com/?grab.php?id=1577&type=wav, or go to http://soundbible.com.

Figure 8.15   What sound will you use?

c. Find the siren file by using the search box on the website.

d. Open the siren file by clicking on its name.

e. Click on the wav icon to download the file.

3. You can check that the file downloaded properly by typing *ls* in the terminal.

4. Once it's finished, type *aplay siren.wav* into the terminal. You should hear a siren!

BOX SOUND NOT WORKING

My computer monitor doesn't have a speaker on it, so when I typed *aplay siren.wav*, nothing happened. I plugged a speaker into the Raspberry Pi, and nothing happened again. Sometimes we need to set where the sound is coming from. Use this command in the terminal

```
amixer cset numid=3 2
```

to set the speaker to the HDMI (monitor) cable. Use

```
amixer cset numid=3 1
```

to set the speaker to the audio jack.

FIND YOUR OWN SOUND

1. Find a sound you like.

2. Look at the number in the address of the sound. My siren was here: http://soundbible.com/1577-Siren-Noise.html.

3. Enter that number, in this case 1577, into the **wget** command: *wget 'http://soundbible.com/grab.php?id=1577&typ**e=wa**v' -O siren.wav.*

Set Up the Door Sensor

1. Open up Python from the main menu: **Programming > Python 3 (IDLE)**.

2. Here's the code to run the siren in Python 3. Let's put it in a function and call that function to test whether it works.

3. Run the code to check whether it works.

```
1  import os
2
3  def soundTheAlarm():
4      os.system("aplay siren.wav")
5
6  soundTheAlarm()
```

Figure 8.16 Code.

```
1   from gpiozero import Button
2   import os
3
4   door = Button(2)
5
6   def soundTheAlarm():
7       os.system("aplay siren.wav")
8
9
10  soundTheAlarm()
```

Figure 8.17 Code.

4. Set up the door as a button using the gpiozero library.
5. When the door button is released, the alarm sounds!

We don't want to drag a TV into our bedroom to get this code running. Let's set up the Raspberry Pi to run this file when it starts, without a screen.

```
1  from gpiozero import Button
2  from signal import pause
3  import os
4
5  #setup the door
6  door = Button(2)
7
8  def soundTheAlarm():
9      os.system("aplay siren.wav")
10
11
12  #when the door is open - sound the alarm!
13  door.when_released = soundTheAlarm
14  pause()
```

Figure 8.18 Code.

CRON

Cron is a tool that lets us set up commands to happen when we want. We can set up Python files to run every hour, every minute, at 12 noon every day, or, in this case, when the Raspberry Pi starts. Find out more about Cron on the Raspberry Pi website (https://www.raspberrypi.org/documentation/linux/usage/cron.md).

1. Open the terminal, and type *crontab –e*.
2. The first time it might ask you to choose an editor. I like nano, so I pressed 2 and ENTER.
3. Press the DOWN key to scroll down to the bottom of the file.
4. On the last line, type your command: *@reboot python3 /home/pi/doorSensor.py*.

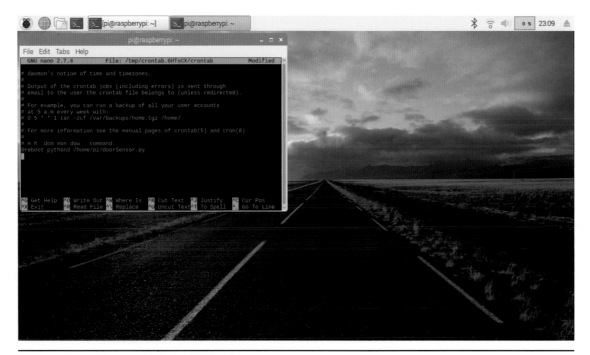

Figure 8.19 Setting up crontab in the terminal.

5. *doorSensor.py* is what I called my program, which we coded earlier. Make sure that you enter your filename correctly here.

6. Restart the Raspberry Pi, and see whether it works!

   a. To test, put the tinfoil pieces together and then take them apart.

   b. The siren should play once the pieces of tinfoil are not touching.

Debug

The only debugging I can see for this project is the build. Getting the cables to stay on the tinfoil while I taped it to a door was tricky. Also, my doors have weird gaps in them. No wonder my house is so cold! I had to make extra-thick tinfoil to make sure that the pieces would touch when the door was closed. What are your doors like?

Expert Level

Check out Mission 12, which sends an email every time your door is opened!

A Lock to Protect Your Sword

We want to keep our sword from Mission 5 safe. Unlike the door in your room, the sword can be anywhere at any time. We want to put a lock on the sword, and if anyone tries to move the sword, an alarm will go off, and the sword won't work. This alarm can go on anything that you want to keep safe, not just the sword.

You don't *need* a speaker for this mission, because the lock will stop the sword from working. We're going to use *if* and *else* statements.

ELSE

We briefly saw an *else* back in the Debug section of Circuit Playground Express in Mission 3. It covers everything the *if* doesn't. Remember the algorithm:

```
If Lorraine = hungry then
      Feed Lorraine
```

What do we do if Lorraine is not hungry? That's where *else* comes along.

```
If Lorraine is hungry
      feed Lorraine
else
      save food for later
```

Only one of these actions will be done—which one depends on if Lorraine is hungry.

Algorithm

```
On start
      Turn Lock on
On Left Tilt
      If the Lock is off
            Go Green
```

```
        Else
              Sound the alarm
On Right Tilt
        If the Lock is off
              Go Blue
        Else
              Sound the alarm
On button A pressed
        Create a cool animation
On button B pressed
        Turn the lock on
On button A and B pressed
        Turn the lock off
```

micro:bit

You don't need headphones to create this mission. The sword will not light up if the lock is on. The headphones act as an extra alarm to let you know that someone is trying to steal your sword.

Build

If you want an alarm, you need to combine the instructions from Missions 5 and 8 to create a sword with a speaker. Instead of a speaker, I'm going to add some headphones.

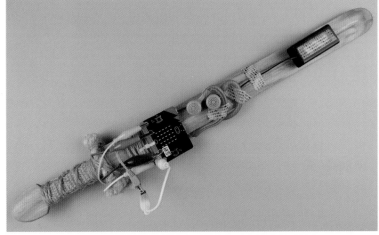

Figure 9.1   Sword lock with headphones attached.

1. Re-create the sword setup from Mission 5.
2. Move the data crocodile clip (green) from pin 0 to pin 1 on the micro:bit. The headphone crocodile clips need to go on pin 0 and GND.
3. Add a crocodile clip to pin 0 on the micro:bit (yellow crocodile clip).
4. Clip the other end of the pin 0 clip onto the tip of the headphone jack.
5. Add a crocodile clip to GND on the micro:bit. You will need to clip it onto the crocodile clip that's already there (white crocodile clip).

6. Clip the other end of the GND clip onto the base of the headphone jack.
7. Secure the headphones to the sword. I wrapped them around the sword, and stuck the earbuds on above the micro:bit.

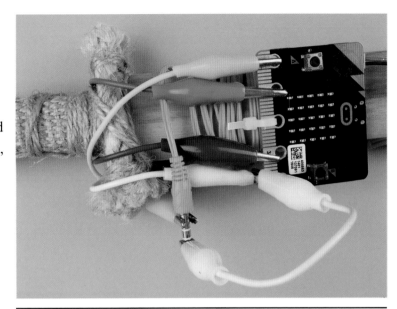

Figure 9.2 Lots of wires!

It's all a bit squishy, but it works!

Going across the micro:bit from pin 0 to GND:

■ Pin 0: yellow crocodile clip connected to the base of the pink headphone jack
■ Pin 1: green crocodile clip connected to the data pin on the light strip
■ Pin 2: a yellow cable tie holding the micro:bit onto the sword
■ 3V: blue crocodile clip connected to 5V pin on the light strip
■ GND: two white crocodile clips, one connected to the light strip and the other connected to the tip of the headphone jack

Code

We're going to tweak the sword code from Mission 6.

1. Change *P0* to *P1* in *set strip to NeoPixel at pin P0 with 8 leds as RGB (GRB format)*.
2. Let's set up the lock and get it locked and unlocked. We're going to create a variable to do this. Learn more about variables in Mission 4.
3. When you press button B, set **Lock** to *0*.
4. From the **Logic** menu, scroll down and drag out the block *true* and place it on top of the 0.

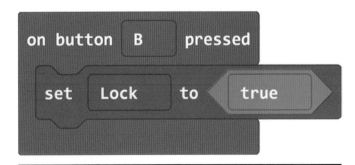

Figure 9.3 Set Lock to true.

BOOLEAN

Variables can be different types of data. Previously, we've used numbers for our variables. In this mission, the variable **Lock** is a *Boolean*. It can only be true or false. There are no other values. This keeps it really simple.

---

5. Create the unlock code: *set Lock to false* under *on button A+B pressed*.
6. Add the code *set Lock to true* to *on start*.
   a. When the micro:bit starts, the lock is on: it is true.
   b. When you press B, the lock is on: it is true.
   c. When you press A and B, the lock is off: it is false.

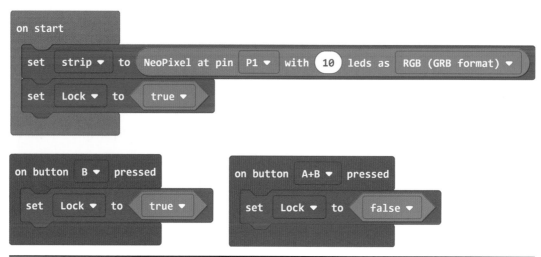

Figure 9.4   Lock on start.

7. Let's work on disabling the tilts if the lock is on.
8. From the **Logic** menu, create the *if* statement around *strip show color blue*.
9. If you have attached the headphones, let's set them off.
10. Click on the plus (+) symbol to add an *else* section. This is where your alarm will go.
11. From the **Music** menu, drag out *start melody dadadum repeating once*, and place it inside the *else*.
12. Do the same for *tilt left*.
13. I like to add an icon as well as an alarm, just in case the headphones don't work.

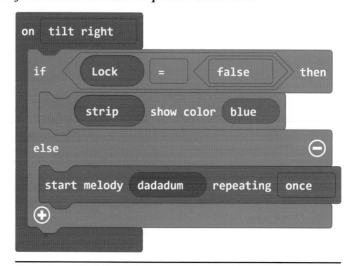

Figure 9.5   Set off the alarm.

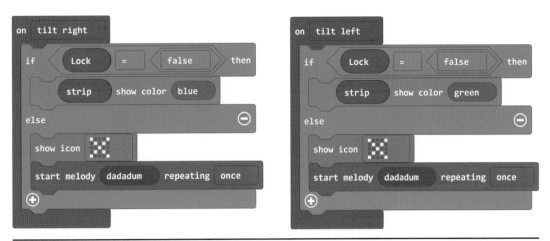

Figure 9.6 Turning off left and right.

14. This is all the code together (except the funky animation code):

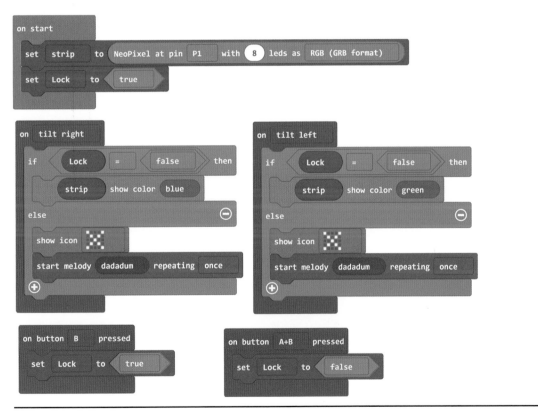

Figure 9.7 Full code.

15. Test the code in the simulator before you download it to the micro:bit.

Now your sword is safe. Don't give your unlock code to anyone!

Debug

Don't forget you need to change the light from P0 to P1 on the micro:bit. This bug caught me out.

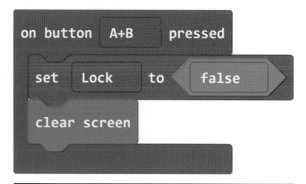

    After playing with my sword for a while, I noticed that if the sword is locked and I try to tilt it, I get the X symbol. When I unlock and tilt, I get a colored sword, but the X symbol is still there.

    We just need to clear the screen when lock changes to false.

Figure 9.8   Clear the screen.

Expert Level

Have a look at Mission 11. It creates a number lock for a cookie jar. Could you add this to your sword lock? Also, in Mission 12, you use the radio to send an alarm to yourself in another room. Instead of a lock, you could use the lock code as a second user profile. User one has the colors blue and green; user two has the colors red and purple.

Circuit Playground Express

Build

You could add some headphones to this build, but they're not much louder than the built-in speaker on the Circuit Playground Express. Follow the instructions in Mission 5 to build your sword.

Figure 9.9   Keeping your sword safe.

Code

Load the code from Mission 5 (we'll need this). Let's set up the lock and get it locked and unlocked. We're going to create a variable to do this. Learn more about variables in Mission 4. *Note:* I'm going to leave out the button A code just to fit everything on the page. You can leave the code in.

1. When button B is pressed, *set Lock to 0.*
2. From the **Logic** menu, scroll down and drag out the block *true* and place it on top of the 0. Learn about Booleans on page 125 of this mission.
3. Create the unlock code *set Lock to false* under **on button A+B pressed**.

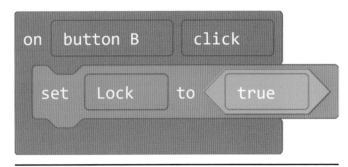

Figure 9.10 Set Lock to true.

4. We want the sword to be locked when we start up. Let's add the code *set Lock to true* to *on start*.

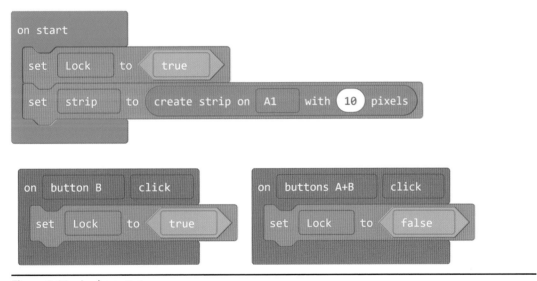

Figure 9.11 Lock on start.

  a. When the Circuit Playground Express starts, the lock is on: it is true.
  b. When you press B, the lock is on: it is true.
  c. When you press A and B, the lock is off: it is false.
5. Let's work on disabling the tilts if the lock is on.
6. From the **Logic** menu, create the *if* statement around *strip set all pixels to blue*.
7. Click on the plus (+) symbol to add an *else* section. This is where your alarm will go.

8. From the **Music** menu, drag out *play sound siren*, and place it inside the *else*.
9. Do the same for *tilt left*.
10. I like to turn on the lights as well as the alarm so that I know the sword is locked.

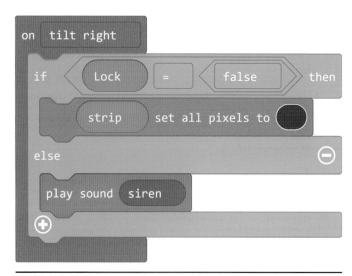

Figure 9.12   Set off the alarm.

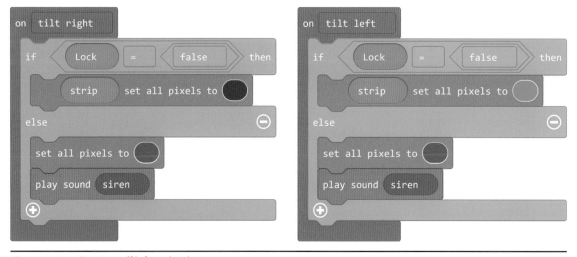

Figure 9.13   Turning off left and right.

This is all the code together:

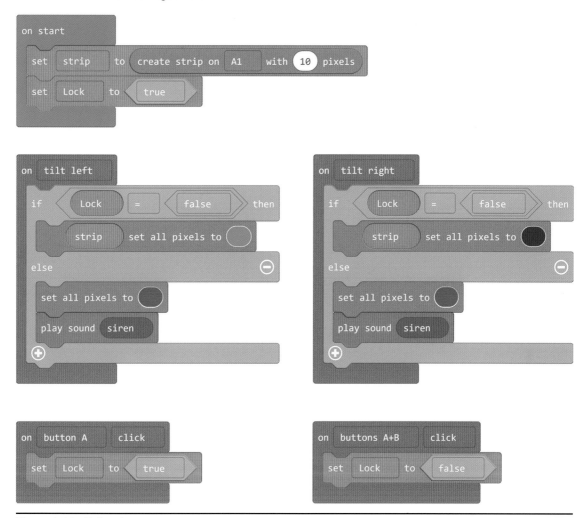

Figure 9.14 Full code.

Now your sword is safe. Don't give your unlock code to anyone!

Debug

When you unlock the sword, the lights stay red. We need to clear the screen when **Lock** changes to *true*.

Expert Level

In Mission 11, you will create a number lock for a cookie jar. You have to enter a number to unlock the jar, not just press A + B. This makes it more secure. Could you add that

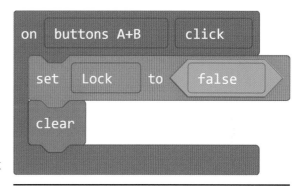

Figure 9.15 Clear the screen.

code to your sword lock? In Mission 12, you will use the radio to send an alarm to yourself in another room. This could be really useful for the sword.

Instead of a lock, you could use the lock code to change to a second user. User one has the colors blue and green; user two has the colors red and purple.

Cookie Jar Protector

Cookies are important; cookies are precious. Let's create an alarm to keep them safe.

You could use the same code and setup from Mission 8 here. What I find, though, is that people spot the tinfoil and know that the alarm is there. In this mission, we create an alarm that's invisible until it's too late, and the perps are caught!

I keep my cookies in a solid jar, none of that see-through nonsense. "Out of sight, out of mind" is my motto when it comes to guarding cookies. This means that when the cookies are safe, it's dark. When someone has opened the cookie lid, it's bright. We're also going to add an extra level of security with each device: a compass sensor for the micro:bit and a sound level sensor for the Circuit Playground Express.

This is how we'll create our alarm.

Algorithm

```
If light sensor detects brightness OR the compass is not pointing North/
    the sound is not loud then
        Sound the alarm
```

OR
In Mission 6, we used an **AND** statement in our algorithm. We were making sure that both button A **and** button B were not pressed. This time we're using an **OR** statement. If the light is bright **OR** the compass is not pointing north/sound is detected, then sound the alarm. We don't want both to have to happen for the alarm to go off. Just one. That's a really important difference between AND and OR.

Build

Find a cookie jar that is wide enough and deep enough for:

- Your cookies
- The micro:bit or the Circuit Playground Express and a battery pack
- Crocodile clips and a set of headphones

The device needs to go on top of the cookies so that the light sensor won't be obscured. The jar cannot be see-through!

Figure 10.1   Cookie jar ready to go.

micro:bit

Code

How Bright Is Bright?

You know what you need to do here: science! We need to know what brightness is to the micro:bit light sensor. Let's show it forever so that when we open the jar, it will change from darkness to bright as we watch the micro:bit (and not the delicious cookies).

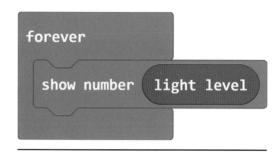

Figure 10.2   What's your brightness number?

Download this code, and place the micro:bit in your jar with the battery attached. Make sure that the micro:bit's lights are facing up, because these are the light sensors. Close the lid. Wait. Open the lid. Wait. What's the number?

My cookie jar is quite deep, so even with the lid off, it's not very bright in there. For my cookie jar, anything above 1 means brightness is detected. If your cookie jar is slightly see-through, your number might be higher than 1.

Test for Brightness

1. Create your selection code inside a *forever* loop.
2. Remember, we need to pause after the melody to allow it to play. This is going to keep playing until the lid is put back on.

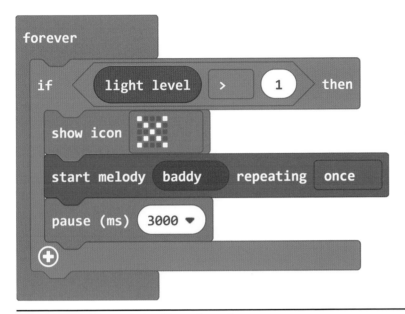

Figure 10.3 What melody will you choose?

Test for Movement

Now let's add our second level of security. Cookie thieves don't always operate in daylight. Sometimes they sneak around in the dark with the lights turned off. Our second level of security will involve the compass that is built into the micro:bit.

1. Again, we need to do some science! The micro:bit in my cookie jar is going to point between 300 and 360 degrees. That means that it will be roughly pointing north/northwest. If it's not pointing toward the north, someone has moved it!

2. Create and download this code to find which way is north from where your cookie jar normally sits. You'll need to place the micro:bit facing in the same direction in the jar every time.

3. Place the micro:bit down flat, and wait a few seconds before you take the reading. Make sure that you're not near something metal, like the fridge, that could throw off the compass.

Figure 10.4 Getting the compass heading.

Note: You might have to calibrate the compass before it will give you a reading. On my micro:bit, it displayed a message "TILT TO FILL SCREEN" before I could continue. I had to tilt the micro:bit until all the dots were filled in. This will happen every time you power up the micro:bit and try to use the compass.

4. Let's code this separate from the light code.

5. I've also added a compass display on the A button so that I can place the micro:bit in the right spot.

6. Download the code, and test it. How much do you need to move the cookie jar before the alarm goes off? Do you need to change the number 300?

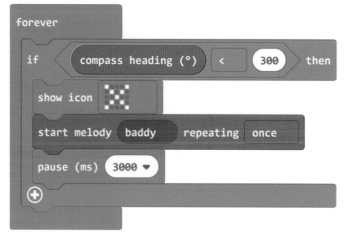

Figure 10.5   Checking compass movement.

Double Security

1. We've got our two tests, so let's combine them. Now we want the alarm to go off if the lid is opened **or** the jar is moved. Not both. We need an **or** block between our two tests.

2. Now place your selections inside the diamonds.

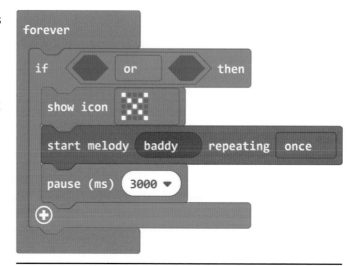

Figure 10.6   OR block.

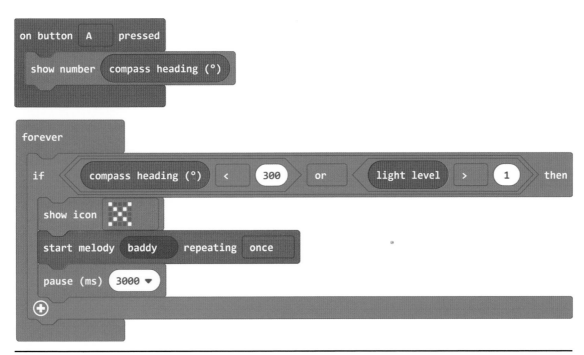

Figure 10.7 Full code.

Debug

As always, we're missing our *clear* block!

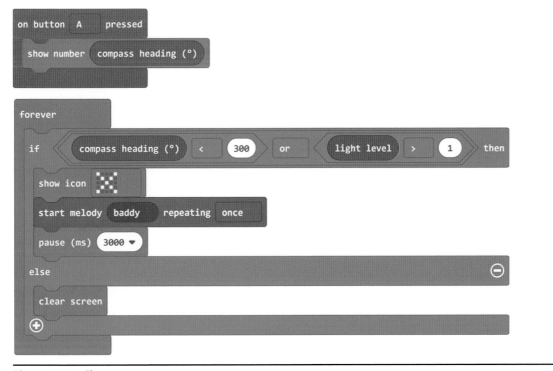

Figure 10.8 Clear screen.

If you use my numbers above, you might find that this sensor doesn't work for your cookie jar in your location. You need to do the experiments and work out the best numbers for your jar in your location. This can be frustrating, especially if you're working in low light.

Expert Level

In Mission 11, we will add a number lock to the cookie jar to turn off the alarm. There's also the radio alarm in Mission 12. Separate from these, you could add more sensors. The alarm will go off if *any* of these conditions are true:

- If the compass heading is less than 300 degrees
- If the brightness is greater than 1
- If the cookie jar is tilted left or right
- If the lid is opened (use the tinfoil from Mission 8)

It will be the most secure cookie jar in the world!

Circuit Playground Express

Code

How Bright Is Bright?

You know what you need to do here: science! We need to know what brightness is to the Circuit Playground Express light sensor.

1. In the **Light** menu, there's a *graph* block that lets us show the light level in lights.
2. The *light level* block came from the **Input** menu.
3. Download the code, and place the Circuit Playground Express in the cookie jar. Press the A button and wait.
4. Run the experiment again. This time put the lid on quickly after you press A.

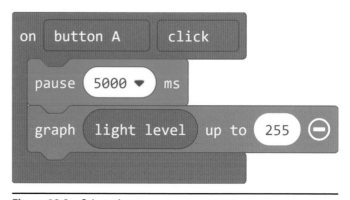

Figure 10.9 Science!

Here are my results on a bright sunny day in Yorkshire:

Lid Off	Lid On
9 of 10 lights = 229.5	0 lights = 0?

Zero is not a great number to work from. I'm going to rerun the experiment with the graph set to 50.

Lid Off (Graph Set to 50)	Lid On (Graph Set to 50)
10 of 10 lights = 50	3 lights = 15

Test for Brightness

Now we're ready to create the light sensor code. Create your selection code inside a *forever* loop.

Test for Laughing

Now let's add our second level of security. Thieves like to cackle loudly as they steal my cookies. Let's use the sound sensor on the Circuit Playground Express to catch them.

Again, we need to do some science! How loud is a cackling laugh? Clear the code for the light level and test the sound level on its own. It's a good idea to test these separately so the data doesn't interfere with each other. Let's graph it out of 255.

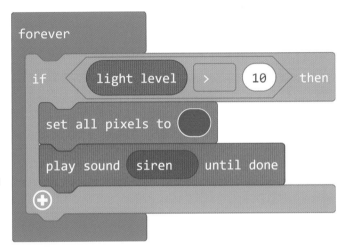

Figure 10.10   What will your alarm sound like?

Figure 10.11   How loud is a cackling laugh?

This is what I got testing with my own evil laugh:

Silence	Loud Laugh
Between 4 and 5 lights = 102 and 127.5	Between 7 and 8 lights = 178.5 and 204

1. Let's code this separate from the light code.
2. Download the code, and test it. How loud do you need to laugh before the alarm goes off? Do you need to change the number 150?

Double Security

We've got our two tests, so let's combine them. Now we want the alarm to go off if the lid is opened **OR** a laugh is sensed. Not both. We need an **OR** block between our two tests.

Now place your selections inside the diamonds.

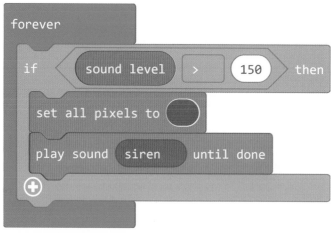

Figure 10.12   Checking for laughter.

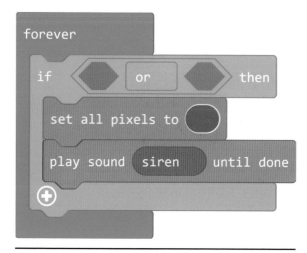

Figure 10.13   OR block.

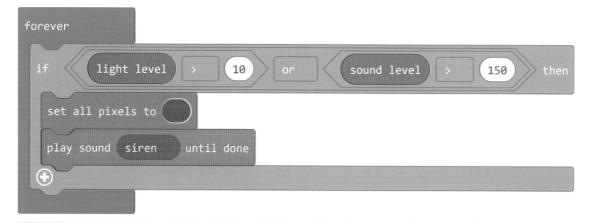

Figure 10.14   Full code.

Debug

As always, we're missing our *clear* block!

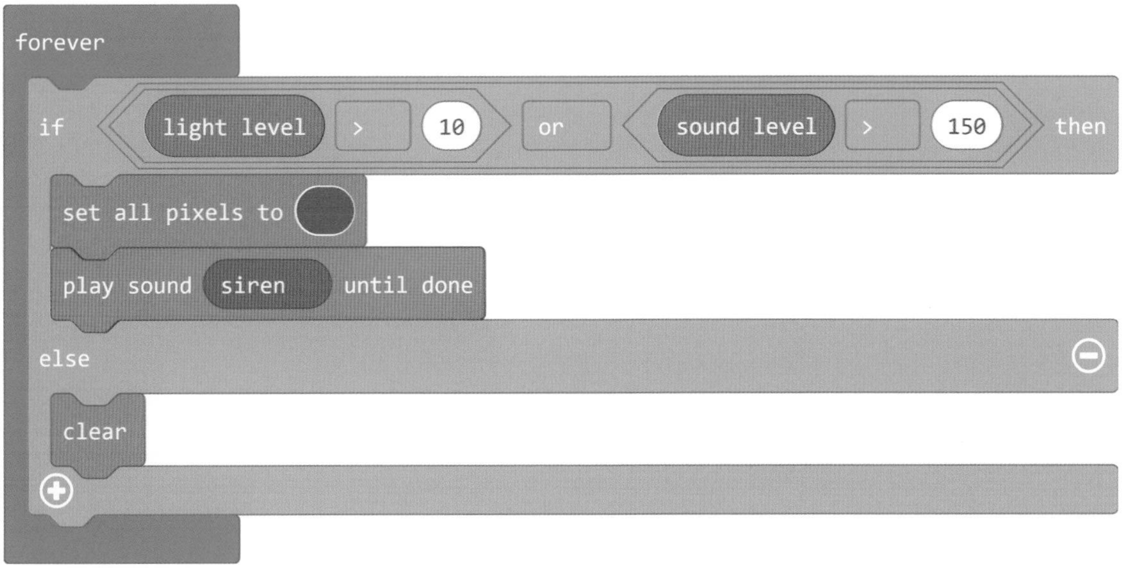

Figure 10.15 Clear screen.

If you use my numbers above, you might find that these security features don't work for your cookie jar in your location. You need to do the science and work out the best numbers for your situation.

Expert Level

In Mission 11, we will add a number lock to the cookie jar to turn off the alarm. There's also the radio alarm in Mission 12. Separate from these, you could add more sensors. The alarm will go off if *any* of these conditions are true:

▪ If the brightness is greater than 1
▪ If the cookie jar is tilted left or right
▪ If the lid is opened (use the tinfoil from Mission 8)

They'll be the most secure cookies in the world!

Number Lock for Your Devices

The problem with the cookie jar alarm is that sometimes we need cookies. And we don't want our cookie enjoyment ruined by an alarm constantly going off! We also need to get into our room without the door alarm going off.

Let's create a number lock for the cookie jar for the micro:bit and Circuit Playground Express and a number lock for the door for the Raspberry Pi. This will be more secure than the sword lock, because you have to enter a **number** to unlock it. The alarm will give you 10 seconds to enter the correct code before it goes off.

We can scroll through numbers using the A button, and submit a guess using the B button on the micro:bit and the Circuit Playground Express. On the Raspberry Pi, we're going to build some real buttons to press.

Algorithm

```
On start
     Lock = True
     Guess = 0
     Passcode = 7
On button A pressed
     Change Guess by 1
     Show Guess (number or light)
On button B pressed
     If Guess = Passcode
          Show a tick/green light
          Turn lock off
     Else
          Show an error/red light
     Set Guess to 0
```

micro:bit

Build

Build the cookie jar from
Mission 10.

Code

Let's create the lock code and then
add it to the cookie jar code later.

Figure 11.1   Cookie jar lock ready to be tested.

Create the Lock Code

1. Create your variables:
   a. *Guess*
   b. *Passcode*
   c. *Lock*
2. Set up your variables in *on start*:
   a. Set *Guess* to *0*.
   b. Set *Passcode* to *7*.
   c. Set *Lock* to *true* from the **Logic** menu.
3. Increase the *Guess* in *on button A pressed*; then show it to the user.

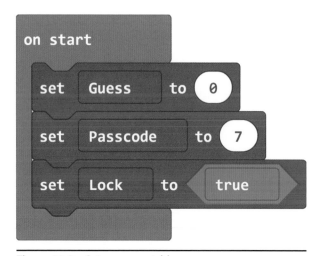

Figure 11.2   Set up our variables.

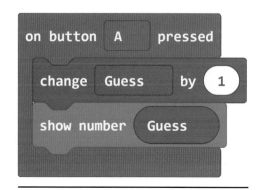

Figure 11.3   Increase Guess.

Let's look back at the algorithm for **on button B pressed**.

```
On button B pressed
    If Guess = Passcode
        Turn lock off
        Show a tick/green light
    Else
        Show an error/red light
    Set Guess to 0
```

Here it is in MakeCode.

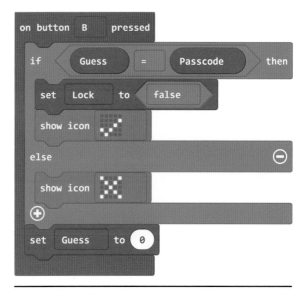

Figure 11.4 B button.

Here are all three pieces of code together:

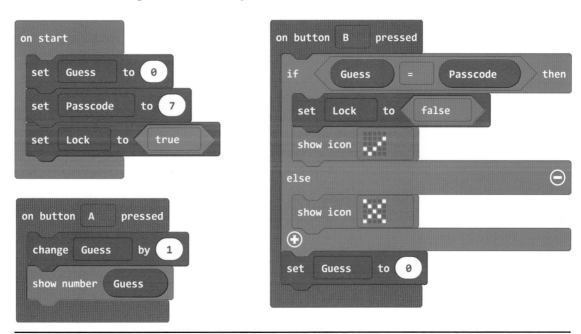

Figure 11.5 Lock code.

Add the Lock to the Cookie Jar

We've got a lock. Let's add this code to the cookie jar code from Mission 10. Here's the cookie jar code:

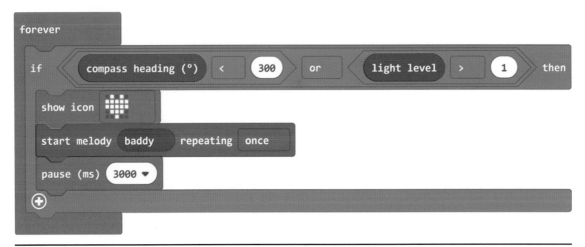

Figure 11.6   Cookie jar code from Mission 10.

Let's work backward and figure out the algorithm: if someone moves or opens the jar, then give him or her 10 seconds to enter the code. If the lock is still on after 10 seconds, the alarm will go off. We need to know whether the jar was moved or opened and when it was moved or opened. In MakeCode, we have a variable that is the amount of time the micro:bit has been running. We can use this to work out whether it has been 10 seconds after the cookie jar was moved or opened.

```
Forever
      If compass heading < 300 or Light level > 1 then
            If BoxOpen = False then
                  Set BoxOpen to True
                  Set BoxOpenTime to the current time
            If the current time >= BoxOpenTime + 10 seconds AND lock is true
                  Sound the alarm
      Else
            BoxOpen = False
```

1. Create the variables ***BoxOpen*** and ***BoxOpenTime***.
2. If the jar was moved or opened, set ***BoxOpen*** to *true* and set the ***running time*** to ***BoxOpenTime***. You'll find the block ***running time (ms)*** under the **Input** menu and then **more**.

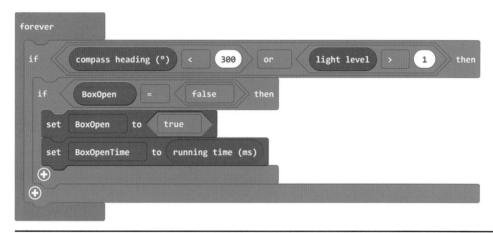

Figure 11.7   Set up the BoxOpen variables.

3. Add an *else* statement here for when the jar is not open or moved by pressing the plus (**+**) symbol on the second *if* block.

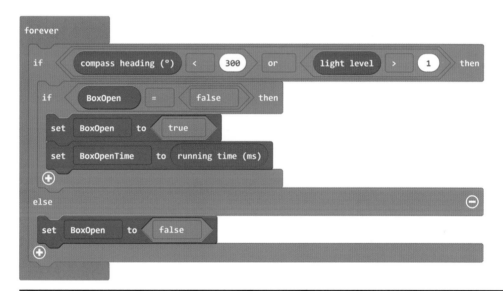

Figure 11.8   An else for when the jar is closed.

4. Now let's see whether the jar was open for more than 10 seconds **AND** the lock is *true*.
5. Build up your *if* statement. Add the ***and*** block first.

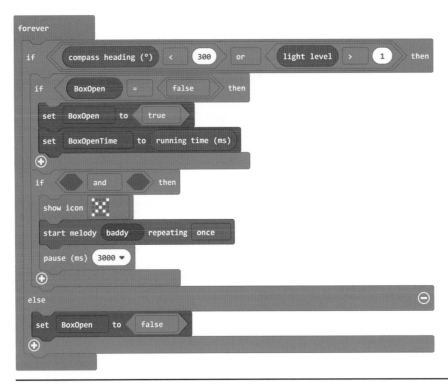

Figure 11.9 Build your if statement.

6. Add *lock = true*.

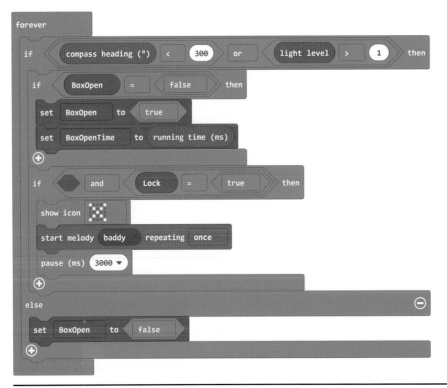

Figure 11.10 Add lock = true.

7. Then check whether the ***running time*** is greater than or equal to ***BoxOpenTime + 10000 ms***. I built this set of blocks using the **Math** menu:

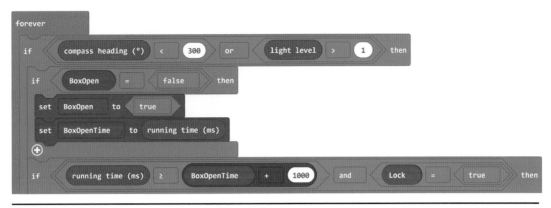

Figure 11.11 Add running time.

And that's it. When the micro:bit starts, the lock is *true*, so the alarm is on. If you enter the right code, the alarm will be turned off, and you can eat your cookies in peace.

Here's all the code together:

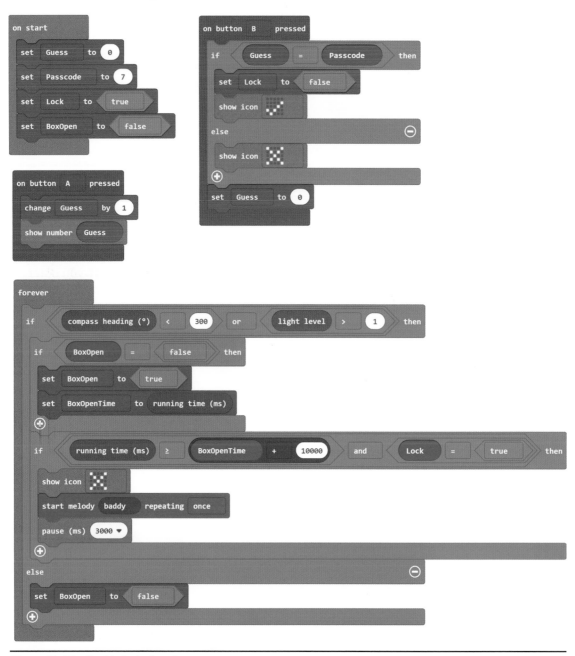

Figure 11.12 Cookie jar lock code.

Did this code work for you? You know what's coming—bugs!

Debug

Before I showed you all the code together in Figure 11.12 my alarm was only set to 1 second, 1000 ms. It should be 10,000 ms.

Instead of 10 seconds, the alarm only gives you 1 second. You need to change the *1000* in the **forever** block to *10000*.

Figure 11.13   Just a small fix.

Expert Level

Try out Mission 12, where you send a radio message to another micro:bit. Try sending a message if someone enters a wrong code. How about turning the lock back on after you've finished your delicious cookie?

Circuit Playground Express

Build

Put together the cookie jar from Mission 10.

Code

Let's create the lock code and then add it to the cookie jar code later.

Create the Lock Code

1. Create your variables:
   a. *Guess*
   b. *Passcode*
   c. *Lock*
2. Set up your variables in *on start*:
   a. Set *Guess* to *0*.
   b. Set *Passcode* to *7*.
   c. Set *Lock* to *true* from the **Logic** menu.

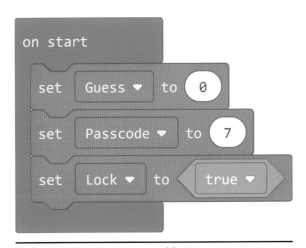

Figure 11.14   Set up our variables.

3. Increase the *Guess* in *on button A pressed*, and then show it to the user using the *graph* block under the **Light** menu.

4. Let's look back at the algorithm for *on button B pressed*:

```
On button B pressed
        If Guess = Passcode
                Turn lock off
                Show a tick/green light
        Else
                Show an error/red light
        Set Guess to 0
```

5. And create it in code.

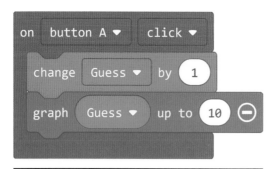

Figure 11.15   Increase Guess.

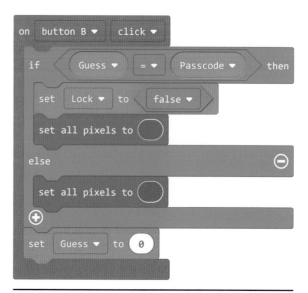

Figure 11.16   B button.

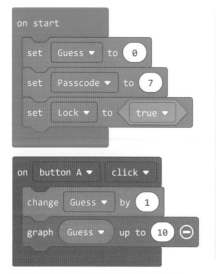

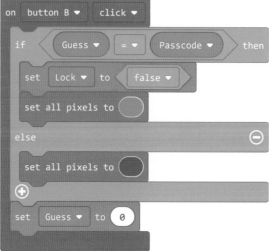

Figure 11.17   Lock code.

This would be a good point to download the code and test it without the cookie jar code.

Add the Lock to the Cookie Jar

We've got a lock. Let's add this code to the cookie jar code from Mission 9. Here's the cookie jar code:

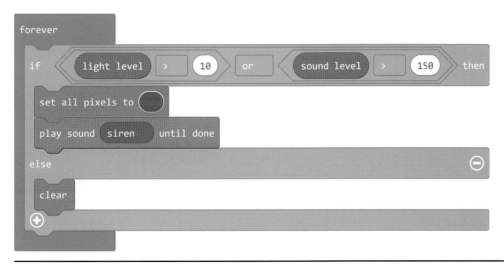

Figure 11.18   Cookie jar code from Mission 10.

Let's work backward and figure out the algorithm: If someone shouts or opens the jar, then give him or her 10 seconds to enter the code. If the lock is still on after 10 seconds, the alarm will go off. We need to know whether the jar has been opened and when it was opened. In MakeCode, we have a variable that is the amount of time the Circuit Playground Express has been running. We can use this to work out whether it has been 10 seconds after the cookie jar was opened.

```
Forever
      If Light level > 10 or sound level > 10 then
            If BoxOpen = False then
                  Set BoxOpen to True
                  Set BoxOpenTime to the current time
            If the current time >= BoxOpenTime + 10 seconds AND lock is true
                  Sound the alarm
      Else
            BoxOpen = False
```

1. Create the variables **BoxOpen** and **BoxOpenTime**.
2. If the jar was opened, set **BoxOpen** to *true* and set **BoxOpenTime** to **millis (ms)**. You'll find the block **millis (ms)** under the **Advanced** menu and then **Control**.
3. I've also added an *else* statement here for when the jar has not been opened, by pressing the plus (**+**) symbol on the *if*.

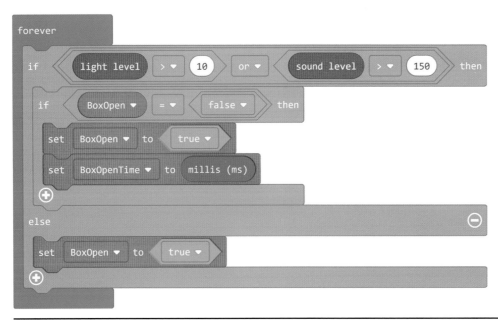

Figure 11.19   An else for when the jar is closed.

1. Now let's see whether the jar was open for more than 10 seconds **AND** the lock is *true*.
2. Build up your *if* statement. Add the ***and*** block first.

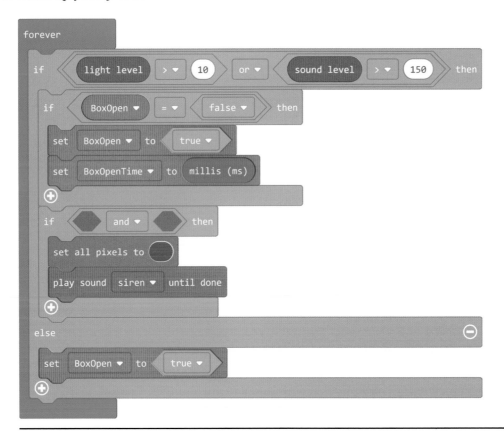

Figure 11.20   Add the and block.

3. Add *lock = true*.

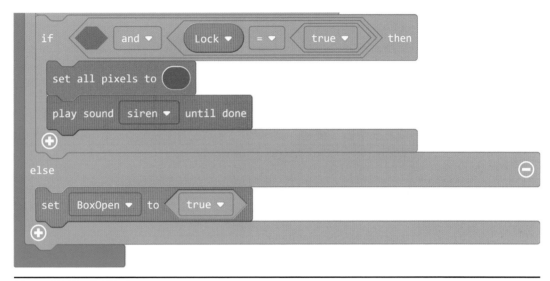

Figure 11.21 Add lock = true.

4. Then check whether the ***millis (ms)*** is greater than or equal to ***BoxOpenTime + 10000 ms***. I built this block using the **Math** menu:

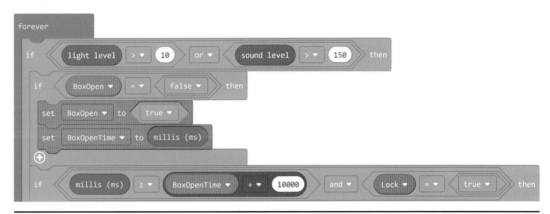

Figure 11.22 Find these blocks under the Math menu.

And that's it. When the Circuit Playground Express starts, the lock is *true*, so the alarm is on. If you enter the right code, the alarm will be turned off, and you can eat your cookies in peace.

Here's all the code together:

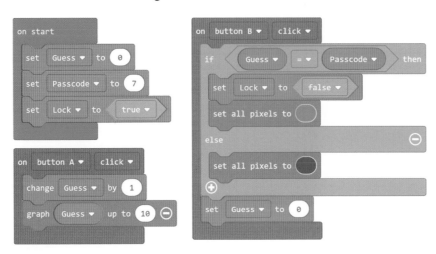

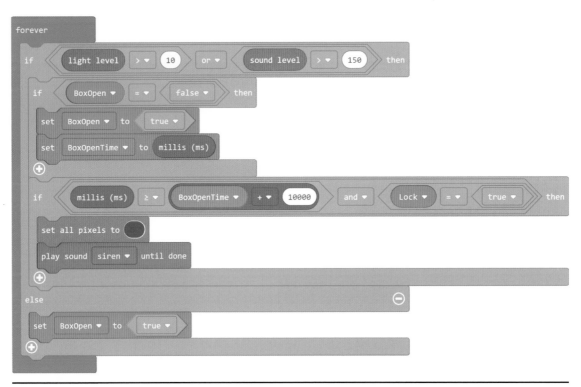

Figure 11.23 Cookie jar lock code.

Did this code work for you? You know what's coming—bugs!

Debug

When you press button A, you can go past the number 10. I think the passcode has to be less than 10, or else you'll take too long to enter it. Plus, it would be difficult to graph the Circuit Playground Express past 10 numbers.

We can stop the user entering a number bigger than 10 using code.

Expert Level

In Mission 12, you will send a radio message to another Circuit Playground Express. How about sending a message when the wrong code is entered? What about turning the lock back on after you've finished your cookie(s)?

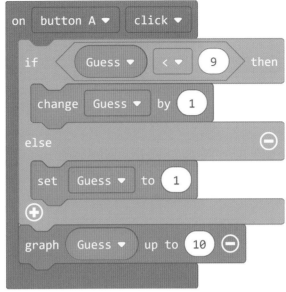

Figure 11.24 Debugging.

Raspberry Pi

Build

For the Raspberry Pi, we're going to create a code lock for the door. In this way, we can enter our door without setting off the alarm.

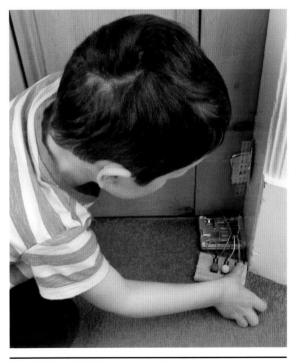

Figure 11.25 Trying to enter the lock code.

You can use arcade buttons or push buttons for this mission. You will need:

■ Three buttons, of different colors

Arcade Button Setup	Push Button Setup
Six male-to-female jumper wires	Four male-to-female jumper wires
Six crocodile clips	Three male-to-male jumper wires
OR	Breadboard
Six male-to-female jumper wires crimped with spade connectors	

■ Two strips of tinfoil
■ Two crocodile clips
■ Two male-to-female jumper wires

The buttons are called *tactile buttons* and are 12 mm in size. I found them in the United States for sale online at Adafruit (https://www.adafruit.com/product/1009) and SparkFun (https://www.sparkfun.com/products/14460). And in the United Kingdom, I found them at the Pi Hut (https://thepihut.com/pages/search-results ?q=tactile) and Pimoroni (https://shop .pimoroni.com/?q=tactile).

Check out Mission 6 on how to connect arcade buttons to the Raspberry Pi. We're setting up three buttons this time. Use pin 18 for the third button.

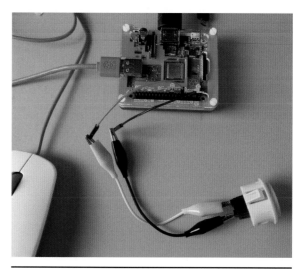

Figure 11.26   One arcade button setup.

For the push buttons setup, we're connecting every button to a GPIO pin and GND. We're going to use a breadboard for this setup.

Let's look at all the equipment we'll need:

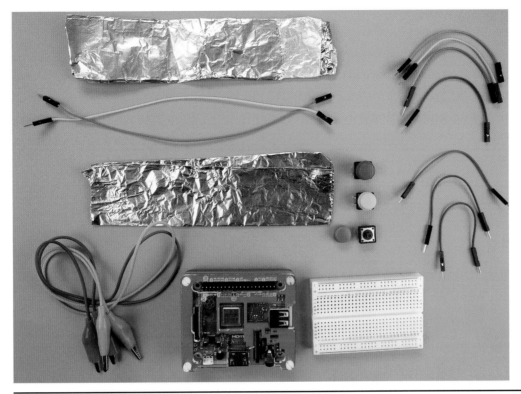

Figure 11.27 Equipment.

The push buttons have lids that come off in different shapes and sizes. I've used short jumper wires for the breadboard and long ones for the door.

1. Clip the three buttons onto the breadboard. Make sure that all four legs are in holes.
2. For each button, put a male-to-male jumper wire in the hole in front of a leg.
3. Put the other end of the male-to-male jumper wires in the ground rail on the edge of the breadboard. It's the blue line marked with a negative symbol (−).
4. This connects all the buttons to the ground rail. Now we can connect the ground rail to ground on the Raspberry Pi. This way we don't use up three ground pins and only use one extra wire for this setup.
5. Use the male end of the four male-to-female jumper wires to:
   a. Connect to a hole in the ground rail.
   b. Connect to a hole in front of the other leg of the red button.

Figure 11.28 Wires.

    c. Connect to a hole in front of the other leg of the yellow button.

    d. Connect to a hole in front of the other leg of the green button.

6. Connect the female ends of the four wires to the Raspberry Pi:

    a. Connect the ground wire (purple) to a ground pin.

    b. Connect the red button wire (orange) to pin 14.

    c. Connect the yellow button wire (yellow) to pin 18.

    d. Connect the green button wire (green) to pin 15.

Figure 11.29   Connected to the Raspberry Pi.

TIPS FOR SETTING UP A BREADBOARD

- The buttons do go in a certain way. Look out for the side where there's only one hole between the two legs.
- Using short jumper wires can be helpful here.

7. Plug your Raspberry Pi into the monitor, keyboard, mouse, and power.

8. Use this code to test your buttons.

9. Build the door lock from Mission 8. You'll eventually need a speaker here too, plugged into the Raspberry Pi.

```
from gpiozero import Button

red = Button(14)
green = Button(15)
yellow = Button(18)


while True:
    if red.is_pressed:
        print("red")

    if green.is_pressed:
        print("green")

    if yellow.is_pressed:
        print("yellow")

```

Figure 11.30   Code.

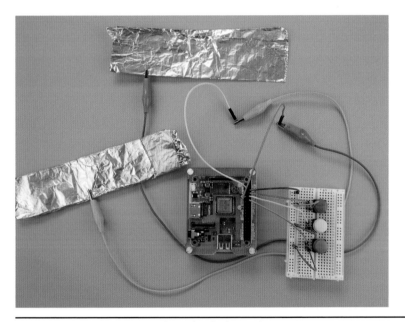

Figure 11.31   Door alarm and number lock together.

We're going to code everything before we move it to the door. For this mission, the TV is not coming with us!

Code

We're not using buttons A and B here, and we don't have any lights, so our algorithm is very different. Let's create a new algorithm. The entry code will be to press the right buttons in the right order. Let's make the entry code three buttons long. We need to check the code after the user has pressed three buttons.

```
Set Guess to empty
Set Passcode = red, red, blue
If button was pressed then
      Add button color to Guess
If three buttons were pressed then
    If Guess = Passcode then
          Play good sound?
    Else
          Sound the alarm
```

We're going to store the passcode in a list.

PYTHON LISTS

A *list* or an *array* in programming is a really useful way of storing multiple variables. Instead of storing **guess1**, **guess2**, and **guess3**, we just create one list called **guess.** This is what the passcode list looks like:

Red	Red	Green

Each variable has its own numbered place in the list *starting from zero*. For example, **passcode[2]** is green. In Python, we can print and edit individual variables in the list or print the whole list. We also can delete and add variables to lists.

---

1. Let's set up the passcode, the buttons, and the empty **guess** list.
2. If a button is pressed, we want to add the name of that color to the **guess** list. You should have also noticed in your test code how one tiny click of a button registers *lots* of button

```
1  from gpiozero import Button
2
3  passcode = ["red", "red", "green"]
4  guess = []
5
6  #setup the buttons
7  red = Button(14)
8  green = Button(18)
9  yellow = Button(15)
```

Figure 11.32   Code.

```
1  print ("ready")
2
3  while True:
4      if red.is_pressed:
5          guess.append("red")
6          sleep(0.3)
7
8      if green.is_pressed:
9          guess.append("green")
10         sleep(0.3)
11
12     if yellow.is_pressed:
13         guess.append("yellow")
14         sleep(0.3)
```

Figure 11.33   Code.

presses. To stop this, I've added a pause after we register a button press. Don't forget to import the time library at the top.

3. After three guesses, let's compare the two lists. For now, let's just print out to the screen "Yes" or "No."

```
1  #they've entered 3 guesses
2  if len(guess) == 3:
3      #do the guesses match the passcode?
4      if guess[0] == passcode[0] and guess[1] == passcode[1] and guess[2] == passcode[2]:
5          print ("Yes, your guess is correct")
6      else:
7          print ("No, your guess is wrong")
```

Figure 11.34 Code.

Now might be a good time to test your code before adding the tinfoil. The code lock is working: the Raspberry Pi knows when you press a button and if you've pressed the right buttons in the right order. We now need to add this code to the door code to get it all working together.

```
Set Guess to empty
Set Passcode = red, red, blue
Setup the buttons
Setup the door
Set When the door is opened: the alarm will sound
If button was pressed then
      Add button color to Guess
If three buttons were pressed then
      If Guess = Passcode then
            Set When the door is opened: nothing happens
      Else
            Set When the door is opened: the alarm will sound
```

1. I've just added six lines to this code:
   a. The function *soundTheAlarm* (don't forget to import the os library)
   b. Setting up the door: *door = Button(2)*
   c. Setting what happens when the door is opened: *door.when_released = soundTheAlarm*
   d. And then inside the *if* statement:
      ■ *door.when_released = None* when the alarm is off.
      ■ *door.when_released = soundTheAlarm* when the alarm is on.

```
1  from gpiozero import Button
2  from time import sleep
3  import os
4
5  passcode = ["red", "red", "green"]
6  guess = []
7
8  #setup the buttons
9  red = Button(14)
10 green = Button(18)
11 yellow = Button(15)
12 #and the door
13 door = Button(2)
14
15 def soundTheAlarm():
16     os.system("aplay siren.wav")
17
18 #set it up: when the door is opened the alarm will sound
19 door.when_released = soundTheAlarm
20
21 print ("ready")
22
23 while True:
24     if red.is_pressed:
25         guess.append("red")
26         sleep(0.3)
27
28     if green.is_pressed:
29         guess.append("green")
30         sleep(0.3)
31
32     if yellow.is_pressed:
33         guess.append("yellow")
34         sleep(0.3)
35
36     #they've entered 3 guesses
37     if len(guess) == 3:
38         #do the guesses match the passcode?
39         if guess[0] == passcode[0] and guess[1] == passcode[1] and guess[2] == passcode[2]:
40
41             #they've guessed the right answer.
42             #when the door is opened, don't sound the alarm
43
44             print ("YES")
45             door.when_released = None
46         else:
47
48             #they've NOT guessed the right answer.
49             #when the door is opened, sound the alarm
50
51             print ("NO")
52             door.when_released = soundTheAlarm
53
54         #set guess back to an empty list
55         guess = []
56
57
```

Figure 11.35   Code.

2. Test this code using tinfoil on your desk next to the monitor. Touch the two pieces of tinfoil together before you run the code.

3. Set up this script to run when you start the Raspberry Pi using Cron instructions from Mission 8.

Debug

While testing this code, I had a lot of problems with the buttons. I added debug messages to check what was happening. I had a print statement after every button press like the code here.

Then I realized that I had mixed up the green and yellow buttons! My photo shows green connecting to GPIO 18 and yellow connecting to GPIO 15, but the code is the opposite.

Debug messages are vital when testing your code. Once you've found your bugs, you can fix them.

It's tricky to know when a button was pressed and registered. Can you play a sound after every button press? Maybe use a short sound file you've downloaded from SoundBible? Check out how to do this in Mission 8.

```
1  while True:
2      if red.is_pressed:
3          print("red")
4          guess.append("red")
5          sleep(0.3)
6
7      if green.is_pressed:
8          print("green")
9          guess.append("green")
10         sleep(0.3)
11
12     if yellow.is_pressed:
13         print("yellow")
14         guess.append("yellow")
15         sleep(0.3)
```

Figure 11.36   Code.

Expert Level

When you turn off the alarm, it's turned off for good. Can you create some code that will turn the alarm back on, either on a different button press or after 5 minutes? If you're not in the house, you're not going to hear the alarm. Send an email when the door has been opened. Check out Mission 12 on how to do this.

Mobile Alarm for Your Devices

Now we can't always be near our room or the cookie jar to keep them safe. We can set up a system where one device sends a message to another one to alert you to a problem.

For the micro:bit and the Circuit Playground Express, we need another device. On the Raspberry Pi, we're going to send an email.

Algorithm

This is the algorithm for the micro:bit and the Circuit Playground Express. For the Raspberry Pi, it's not too different. We're just sending an email on device 1 and receiving it on device 2.

Device 1	Device 2

IF door/cookie jar is opened
    Send number
	On number received
	    Display alarm

micro:bit

Build

Build your cookie jar or door sensor as shown in previous missions. The second micro:bit needs to be in range of the first. You will need to experiment to figure out how far apart your micro:bits can be.

I'd recommend labeling your micro:bits somehow with a sticker to know which is which.

Figure 12.1 His cookies are safe and sound with his micro:bit alarm.

167

Code

Cookie Jar micro:bit

Let's start with the micro:bit that is attached to the cookie jar. Here is the code from Mission 10:

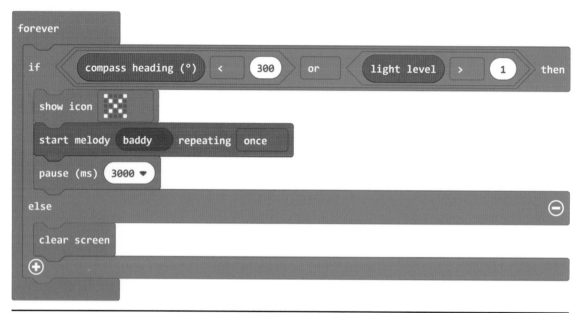

Figure 12.2 Cookie alarm.

We need to set up the radio group under ***on start***.

RADIO GROUPS
Radio Groups make sure that your messages only go to the micro:bit you want it to go to. Keep this number safe from other micro:bit owners nearby.

In the cookie jar code, where the alarm goes off, add the radio code to send a number.

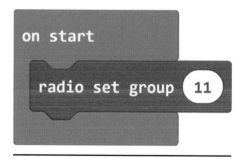

Figure 12.3 Set the group number.

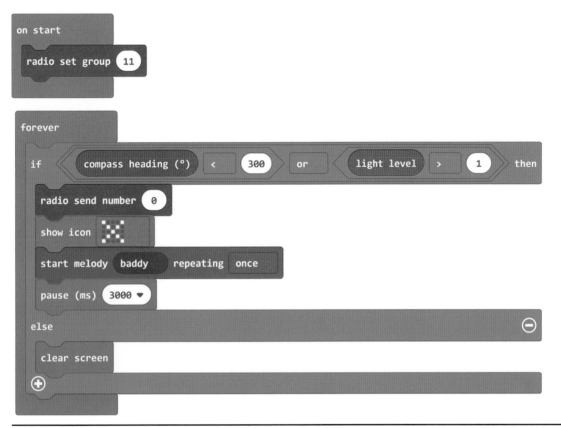

Figure 12.4 Send a number.

That's it for the cookie jar micro:bit. If you don't want an alarm on the jar, you can remove the ***melody*** block.

Door micro:bit

The same goes for the door micro:bit from Mission 8.

1. Add the radio group to ***on start***.
2. Send a number when the alarm is meant to go off.

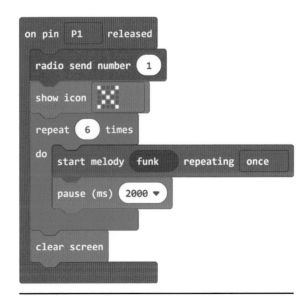

Figure 12.5 Door alarm.

Alarm micro:bit

Now let's set off the alarm on the second micro:bit that doesn't have to be next to the cookie jar or your room door. You could just carry it with you!

I'm learning my lessons here. I've put in an icon in case the speaker disconnects, and I'm clearing the screen after 3 seconds.

If you have three micro:bits, what you could do is have a different tune and letter for each alarm that might go off. Create condition statements to check which alarm went off. To get the variable *receivedNumber* drag it down from the block **on radio received receivedNumber**. The cookie sent the number 0, and the door sent the number 1.

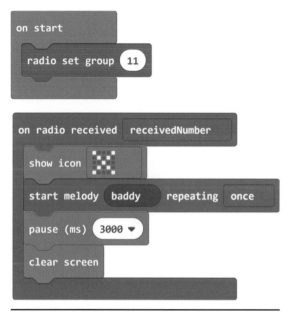

Figure 12.6   Set off the alarm.

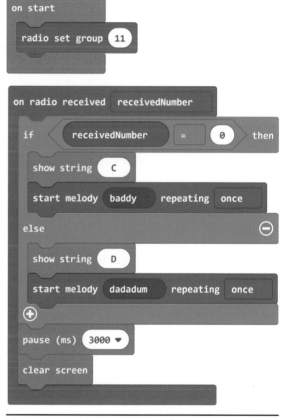

Figure 12.7   What alarm went off?

Debug

My code worked really well the first time. I know a lot of people get the block **set radio group** wrong. This *has* to be in **on start**, or nothing will happen.

Other debugging problems will come with the distance between the two devices. How far away can your alarm be? Will your alarm work on different floors? Will it work through closed doors? Experiment with the alarm to see what the limit of the range of your micro:bit is.

Expert Level

A friend has entered the house. You don't want to give him or her the cookie jar password, but you do want to let him or her have a cookie. Send a message to the cookie jar to turn off the alarm for 5 seconds.

Circuit Playground Express

Build

Build your cookie jar or door sensor as in previous missions. How close does your second Circuit Playground Express need to be? You'll need to experiment.

Figure 12.8   Circuit Playground Express cookie jar ready.

The Circuit Playground Express uses infrared to send messages from itself to the other Circuit Playground Express. Infrared works by line of sight. The two Circuit Playground Expresses need to be in the same room. Think about how far away your remote control for your TV works—this uses the same technology.

Code

Cookie Circuit Playground Express

Let's start with the Circuit Playground Express that is attached to the cookie jar. Here is the code from Mission 10.

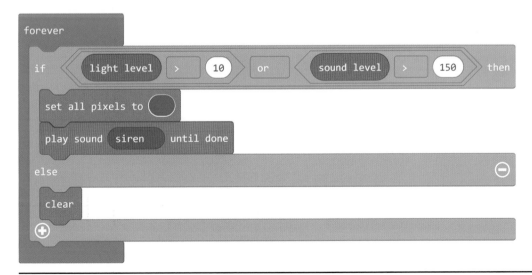

Figure 12.9   Cookie alarm.

In the cookie jar code, where the alarm goes off, add the code to send a number.

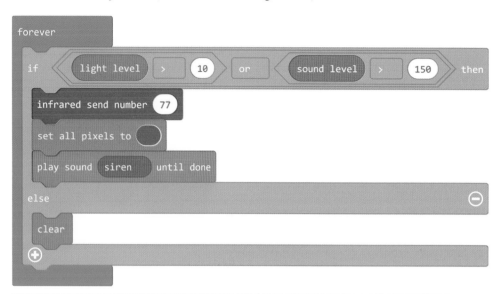

Figure 12.10   Send a number.

That's it for the cookie Circuit Playground Express! If you don't want an alarm on the cookie jar, you can remove the *play sound* block. But you should replace it with a *pause* block so that only one signal is sent every 2 or 3 seconds. You don't want hundreds of signals being sent!

Door Circuit Playground Express

The same goes for the door with the Circuit Playground Express. Send a number when the alarm is meant to go off. And don't forget to set the pin to pull up when the Circuit Playground Express starts.

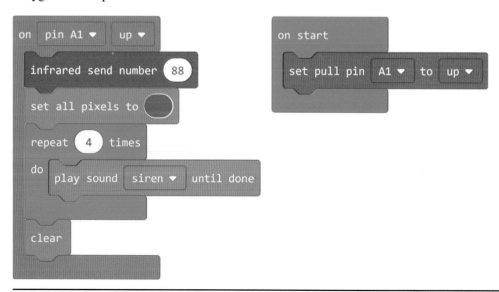

Figure 12.11   Door alarm.

Alarm

Now let's set off the alarm on the second Circuit Playground Express that doesn't have to be next to the cookie jar or your room door. You could just carry it with you!

I'm learning my lessons here. I've changed all the pixels to red in case the speaker disconnects, and I'm clearing the screen after 3 seconds.

If you have three Circuit Playground Expresses, what you could do is have a different tune and color for each alarm that might go off. Create condition statements to check which alarm went off. The cookie jar sent the number 77, and the door sent the number 88.

Debug

This is a physical debug. How far will your alarm go? Do the Circuit Playground Expresses need to be pointing at each other? Experiment with the alarm to see what the limit of the range of your Circuit Playground Express is.

Expert Level

Sometimes I'm lazy and I want someone else to get me a cookie. I don't want to give that person the password so that he or she can steal a cookie every time. I just want to open the jar for 5 seconds. Send a message to the cookie jar to turn off the alarm for 5 seconds.

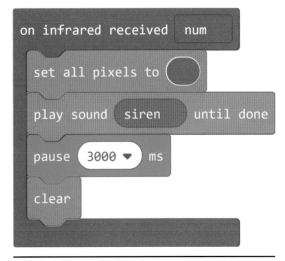

Figure 12.12 Set off the alarm.

Figure 12.13 Which alarm went off?

Raspberry Pi
Build

Follow the build steps in Mission 8 to create the door alarm. The Raspberry Pi needs to be connected to the internet to be able to send an email. Disconnect the Raspberry Pi from the door; plug it into a monitor, keyboard, and mouse to code; and test it.

Code

You will need access to an email address to get this mission working. I've tested this mission with a Gmail account that I set up just for coding projects. I find it's easier to have a separate email address from my main personal email for coding projects. You will need an adult to help you set up and get the app password for a Gmail account.

Figure 12.14 Email alert.

Get the Code from Gmail

1. Sign up for a Google account at https://accounts.google.com/signup.
2. Log into your Gmail account at https://gmail.com/.
3. Select your profile picture in the top right, and select Manage Your Google Account.
4. Select *Security*.
5. Under *Signing in to Google*, select *App passwords*. *Note:* Two-step authentication has to be turned on.
6. Click on *Select app*, and choose *Other* (custom name).
7. Type in a name. I chose "**Door alarm**."
8. Copy down the password you're given.

Create and Test the Python Code for Emailing

1. Here's the Python code to send an email. Change the following:
   a. *From_email* to your email address
   b. *To_email* to your email address
   c. *Password* to the app password you got in the earlier steps

```
 1  from savetheworld import email
 2
 3  #setup the variables
 4  from_email = "youraddress@gmail.com"
 5  to_email = "youraddress@gmail.com"
 6  password = "SECRET"
 7
 8  #create your message
 9  message = """\
10  Subject: Test email!
11  Here's a test email. Did it work?
12  Sent from your Raspberry pi"""
13
14  email.send(from_email, to_email, message, password)
```

Figure 12.15 Setup.

You need the *savetheworld* library to get this code working. Learn how to download it from Mission 1.

PYTHON STRING

The variable *message* is a really long string of text. To see it better in my code, I spread it over several lines. To do this, I used three quotation marks at the start and end of the variable ("""*variable*""").

---

2. This code will send an email to you. Run the code to test it.

Remember that the password is not your normal email password; it's the app password you got from Google. You can use a different email address for the *to_email* variable.

Add to the Door Code

In the door code from Mission 8, we had a function called *soundTheAlarm*. That function played a siren sound. This is where we will put our email code.

1. Don't forget to import the libraries we need to send email and set up the variables:
   a. *gpiozero*
   b. *signal*
   c. *os*
2. Set up the *door* variable.
3. Change your *message* text to something about a door being opened.
4. Add the function *soundTheAlarm* and its code from Mission 8.
5. Place the *email* code in *soundTheAlarm*. I also put a print in here so that I know something has happened when I'm testing it.
6. Call the function *soundTheAlarm* when the door is released.

```
1   from savetheworld import email
2   from gpiozero import Button
3   from signal import pause
4   import os
5
6   #setup the door
7   door = Button(2)
8
9   #setup the email variables
10  from_email = "youraddress@gmail.com"
11  to_email = "youraddress@gmail.com"
12  password = "SECRET"
13
14  #create your message
15  message = """\
16  Subject: Door has been opened again!
17  Your door was opened!
18  Sent from your Raspberry pi"""
19
20  def soundTheAlarm():
21      email.send(from_email, to_email, message, password)
22      print("email sent!")
23
24  door.when_released = soundTheAlarm
25  pause()
```

Figure 12.16 Code.

Be careful with this code. If you start sending hundreds of automatic emails, Gmail will not be happy and might suspend your account. *Only* use your programming powers for good.

Debug

This code just worked like magic for me! If you're having problems, check that your Raspberry Pi is connected to the internet, your email addresses are right, and you've got the app password right. Long passwords can be tricky to type out.

Expert Level

Add the code to Mission 11. If you enter the wrong lock code, send a different email: "Intruder alert!"

Floor Mat Alarm

For rooms where there are no doors or doors on which we don't want to put electronics on the outside of, we can create a floor mat alarm. This is perfect if you share a room with a sibling and don't want him or her on your side of the room. You create a sensor that will set off an alarm if he or she steps on it.

Algorithm

This sensor is basically the opposite of the door and cookie sensors. Rather than breaking the circuit, the circuit is created. It's Mission 1 with a floor mat instead of you!

```
If circuit is created
        Set off alarm
```

What's very different is the build. The first part of the build is the same for all three devices.

Build

The build is the same for all devices, up to the moment you attach the device. We're going to create a "circuit sandwich." The circuit will be complete when the two pieces of bread touch each other. In the middle there will be some filling to stop the bread from touching all the time. There will be a hole in the filling so that the circuits can touch when the sandwich is squished. I'm going to use some foam to create my filling.

What You Will Need

- Two pieces of cardstock
- Tinfoil to cover one side of each of the A4 cards
- Two crocodile clips
- Foam, cut to just bigger than A4 size
- Tape
- Scissors

Build

1. Cover one side of the card in foil.
   a. Cut off a strip of tinfoil that is a bit longer than the A4 piece of card.
   b. Place the card on top of the tinfoil.
   c. Tape down the tinfoil.
   d. Do the same for the other piece of card.
2. Cut a hole in the middle of the piece of foam.
3. Place the card, tinfoil face down, onto each side of the foam.

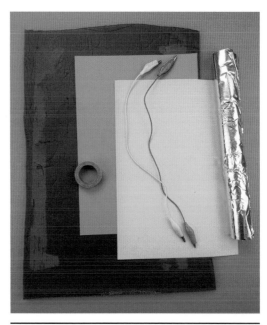

Figure 13.1   Your tools.

Figure 13.2   Card ready.

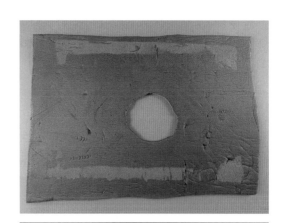

Figure 13.3   Holey foam.

4. I used a strong tape to tape the card onto the mat.
5. Do the same on the other side of the mat.
6. Test your setup by squishing the card down onto the mat. Does it touch the other card through the hole? It can be difficult to tell.
7. Add a crocodile clip to the tinfoil on each side of the mat.

My blue card is on the other side.

Now add this floor mat to your device. Here it is in practice at my front door (I've hidden it under a doormat).

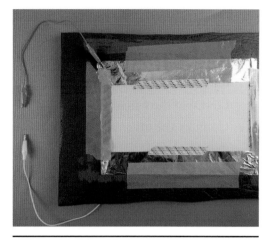

Figure 13.4 Floor mat.

micro:bit

Build

1. Connect one crocodile clip from the mat to ground (GND) on the micro:bit.
2. Connect the other crocodile clip from the mat to pin 1 on the micro:bit.
3. Add the speaker to pin 0 and GND.

Code

Way back in Mission 1, you created the same code for this project with the touch sensor. Add everything we've learned since then into the alarm from Mission 12.

Figure 13.5 Catching intruders.

Expert Level

Sometimes people don't set off the alarm. They step around the sensor, or they are cats and don't weigh enough to press down on the sensor. To scare these smug people, set off the floor mat micro:bit alarm from a second micro:bit.

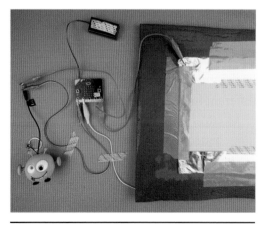

Figure 13.6 micro:bit floor mat.

Circuit Playground Express

Build

1. Connect one crocodile clip from the mat to GND on the Circuit Playground Express.
2. Connect the other crocodile clip from the mat to pin A1 on the Circuit Playground Express.
3. Add the speaker to pin A0 and a different GND.

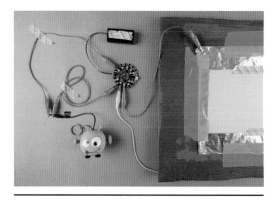

Figure 13.7   Circuit Playground Express floor mat alarm.

Code

Way back in Mission 1, you created the same code for this project with the touch sensor.

Let's add everything we've learned since then into this alarm.

1. We need the pin to pull up on start.
2. Add a speaker to pin A0, and play a tune when the mat is stepped on (Figure 13.9).
3. Send an infrared signal from this Circuit Playground Express to another Circuit Playground Express with the speaker attached (Figure 13.10).
4. You'll find an example of this code in Mission 12 (Figure 12.12).

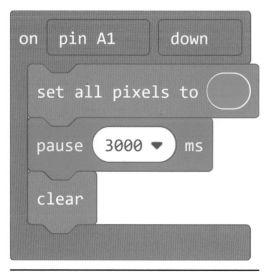

Figure 13.8   Code from Mission 1.

Debug

Did it work? Problems with this mission can come from the build. Your foam might be too thick, or the hole might not be big enough. If your foam isn't thick enough, it won't bounce back, and the cards will continue to touch. You might need to rebuild the mission with different materials if it's not working for you.

Mission 1 was designed to check whether someone was a zombie, so it would show a green light. This is an intruder! We need a red light! Or a flashing red light!

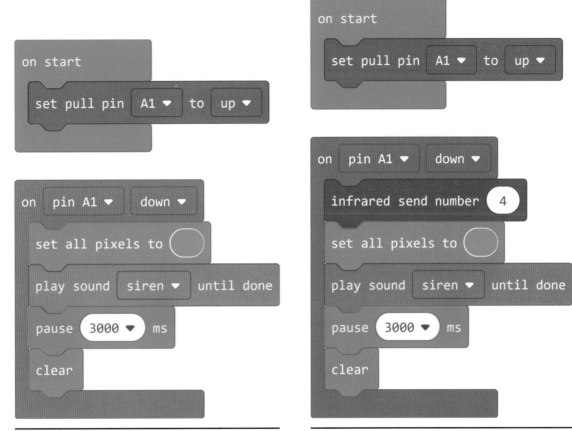

Figure 13.9 Play some tunes.

Figure 13.10 Send a radio message.

Expert Level

I set up this floor alarm trap for my family, and my youngest son stepped on it, smiled at me, and then stepped off. The alarm didn't go off! But it went off for his brother and Dad. Why?? It turns out that he's not that heavy! His weight did not push the cards together through the foam. You might find that it doesn't work for small pets either.

To scare these smug people, set off the floor mat alarm from a second Circuit Playground Express.

Raspberry Pi

Build

This is the same build as the door alarm.

1. Connect one crocodile clip from the mat to the male end of a male-to-female jumper wire.
2. Connect the female end to a ground pin on the Raspberry Pi.

3. Connect the other crocodile clip from the mat to the male end of a male-to-female jumper wire.

4. Connect the female end to a data pin on the Raspberry Pi. I went with GPIO 2.

5. Connect a speaker to the Raspberry Pi.

Code

In Mission 8, we created the door sensor alarm. This went off when the door was opened. We just need to tweak that code to go off when the circuit is closed.

```
1  from gpiozero import Button
2  from signal import pause
3  import os
4
5  #setup the floor matt
6  floor = Button(2)
7
8  def soundTheAlarm():
9      os.system("aplay siren.wav")
10
11 #when the matt is stood on - sound the alarm!
12 floor.when_pressed = soundTheAlarm
13 pause()
```

Figure 13.11 Code.

1. I've changed **door** to *floor*.

2. And I called *when_pressed* instead of *when_released*.

Let's add everything we've learned since then into this alarm.

1. Send an email when the floor mat is pressed.

```
1  from savetheworld import email
2  from gpiozero import Button
3  from signal import pause
4  import os
5
6  #setup the floor matt
7  floor = Button(2)
8
9  #setup the email variables
10 from_email = "youremail@gmail.com"
11 to_email = "youremail@gmail.com"
12 password = "123456789"
13
14 #create your message
15 message = """\
16 Subject: Door has been opened
17 Your door was opened!
18 This message was sent from the Raspberry Pi on your bedroom door"""
19
20 def soundTheAlarm():
21     os.system("aplay siren.wav")
22     print ("alarm")
23     email.send(from_email, to_email, message, password)
24     print("email sent!")
25
26 #when the matt is stood on - sound the alarm!
27 floor.when_pressed = soundTheAlarm
28 pause()
```

Figure 13.12 Code.

2. Remember to change the email addresses and the password to yours as in Mission 12.

3. You're not going to bring the monitor and keyboard to the floor mat. Set up a crontab to run this code when the Raspberry Pi powers up. Find the steps on how to do that in Mission 8.

Debug

As with the door project, the tricky part of this project is the build. You need the foam to be thick enough so that the cards don't touch all the time, but not too thick so that the cards do touch when pressed.

Expert Level

How about adding the number lock from Mission 11 here? When you enter your house, you have to enter the correct code, or the alarm will go off.

Treasure Box Alarm

Like all normal adults, I have a treasure box full of valuables in my room. I need an alarm to keep my treasure safe, but all my previous alarms won't work.

My treasure box is so full that the lid won't close. This means that I can't use the tinfoil alarm or the brightness alarm. My treasure box is also really heavy. No one is going to move it. A thief is just going to open the lid and take my treasure. I can't use the compass alarm.

What I can use is a movement alarm. If the lid is lifted up, it can be sensed and set off an alarm.

Algorithm

```
On movement
        Set off alarm
```

micro:bit

Build

I've taped my micro:bit inside the treasure box lid so that to the outside world it's invisible. The battery is also taped inside.

Code

Find Z

Oh yes, it's science time! The micro:bit measures acceleration using three axes: X, Y, and Z. If the micro:bit is lying flat on the table, X and Y should

Figure 14.1   Measuring inside the lid.

be 0, and Z will be around −1,000. This is because gravity is pulling on the micro:bit. If you lift up the front of the micro:bit, Z and Y will change. If you pick up one of the sides, Z and X will change. Z is the only point that always changes with movement. Let's measure Z.

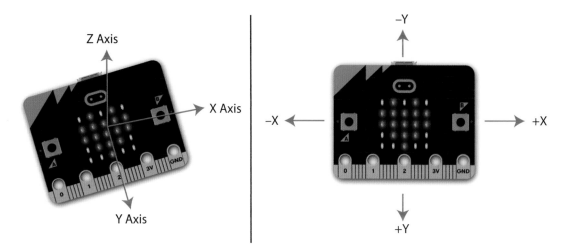

We need to know what Z is when the lid is closed. Let's record Z in a variable that we can look at later.

1. Download this code to the micro:bit, and place it inside the treasure box lid.
2. Press A, and then close the lid.
3. Wait a few seconds, and open the lid.
4. Press B to show the value of Z when the lid was closed.
5. Repeat the experiment when the lid is open.

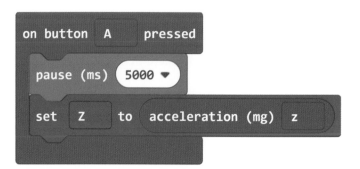

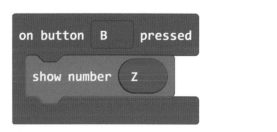

Figure 14.2 Test for Z.

My micro:bit showed 1,024 when the lid was closed, and it showed −1,008 when the lid was fully open. That's a huge range.

Let's see what Z is when the lid is halfway open. Press A, half close the lid, keep it steady for 5 seconds, wait a bit longer, and then open it and press B. I got 992.

Fully open: −1,008
Fully closed: 1,024
Halfway open: 992

Checking Z

If I say *if Z is less than 1,000*, that should catch anyone opening the treasure box, even the tiniest bit.

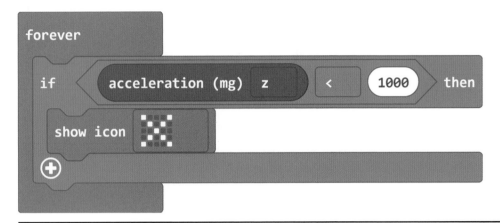

Figure 14.3 Checking for thieves.

Let's add a speaker and another micro:bit alarm.

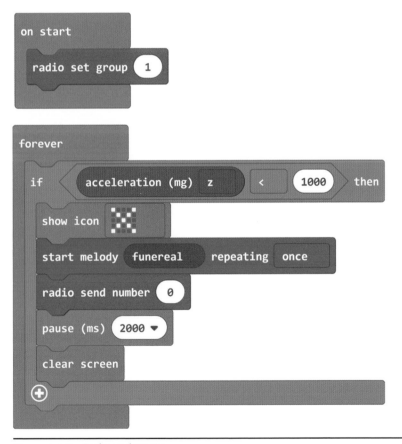

Figure 14.4 Radio and music.

And the second micro:bit can alert us to our treasure being touched. Remember, this code is in Mission 12 (Figure 12.6).

Debug

I left a bug in the preceding code on purpose. Honest, I did. The radio group on the micro:bit inside the treasure box is 1. The radio group on the micro:bit outside is 33. These micro:bits will not talk to each other! To do that, they need to be in the same group.

My debugging on this mission all happened in the build stage. It took me several attempts to get the range right. Sometimes you have to tweak your code, download, test, tweak again, download, and test again to get it working in live conditions. This can be a slow process.

Expert Level

How many times a day is your treasure box opened? Record an opening in a variable, and display that variable on a button press

Circuit Playground Express

Build

I've taped the Circuit Playground Express inside the treasure lid so that it's invisible to the outside world. The battery is also taped inside.

Code

Find Z

Time for some experiments! The Circuit Playground Express measures acceleration using three axes: X, Y, and Z. If the Circuit Playground Express is lying flat, X and Y should be 0, and Z will be around $-1,000$. This is because gravity is pulling on the Circuit Playground Express. If you lift up the front of the Circuit Playground Express, Z and Y will change. If you pick up one of the sides, Z and X will change. Z is the one axis that changes any time there's movement. Let's measure Z.

We need to know what Z is when the lid is closed. Let's record Z in a variable that we can look at later.

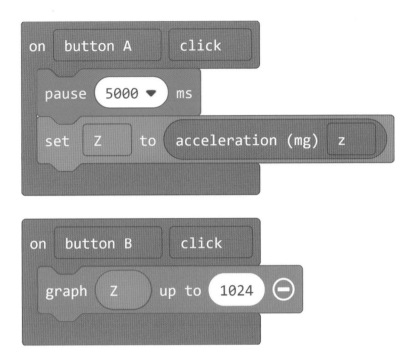

Figure 14.5 Test for Z.

1. Download this code to the Circuit Playground Express, and place it inside the treasure box lid.
2. Press A, and then close the lid.
3. Wait a few seconds, and open the lid.
4. Press B to get the value of Z when the lid was closed.
5. Repeat the experiment when the lid is open.

My Circuit Playground Express showed all ten lights on when the lid was closed. Each light represents 102.4. My Circuit Playground Express showed four lights on when the lid was fully open. Four lights means $102.4 \times 4 = 409.6$.

Fully closed = 1,024
Fully open = 409.6

Checking Z

If I say *if Z is less than 900*, that should catch anyone opening the treasure box, even the tiniest bit.

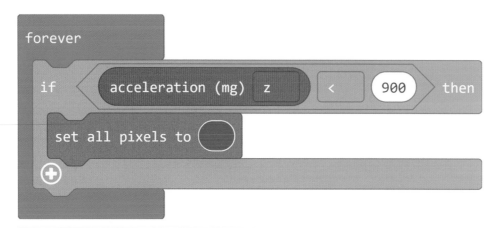

Figure 14.6   Checking for thieves.

Let's add a speaker and another Circuit Playground Express alarm.

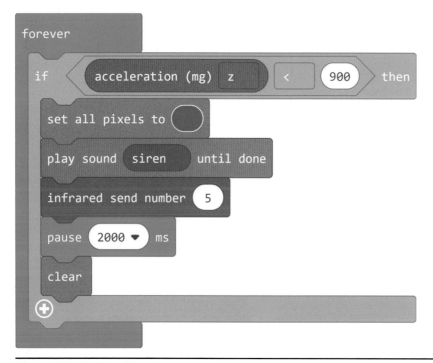

Figure 14.7   Radio and music.

And the second Circuit Playground Express can alert us to our treasure being touched. Remember that this code is in Mission 12 (Figure 12.12).

Debug

The alarm sounds on the treasure box and then sends a message to the other Circuit Playground Express. You might want to send the message first, because it takes a few seconds for the siren to finish playing.

When you're coding projects that you also have to build, it's difficult to know where the bug is. It took me several attempts to get the number 900 right. I had it set too low and then too high! I had to change the code, download it, and put the Circuit Playground Express back in the treasure box. This was tiresome, but now it works.

Expert Level

Like the door alarm, record how many times a day your treasure box has been opened in a variable. Display that variable on a button press. When you leave the house in the morning, open the treasure box so that the number will be 1. When you get home, is the number 2 or is it 1?? If it's 1, someone opened the treasure box and then reset the Circuit Playground Express to try to get away with it.

PART THREE

Save the World

Now that you've kept zombies away and looked after your family
and possessions, it's time to save the world.

Step Counter

Our first mission in saving the planet is to create a step counter. Walking is great for both the environment and you. No smoke comes out of you when you walk, and you don't take up resources such as gasoline or diesel fuel to move. Let's make walking more fun and competitive with a step counter.

Algorithm

```
On move
        Increase step by 1
        Display a graph of steps
```

micro:bit

Build

We want to attach the micro:bit to our wrist in a comfortable way. I found some wrist sweatbands with zips in them that are perfect for holding a micro:bit and a battery. With this black one, you can even see the numbers on the micro:bit through the material.

Alternatively, the MakeCode team from Microsoft has a great tutorial on how to create a micro:bit holder for your watch using material from an old T-shirt (https://makecode.microbit.org/projects/watch).

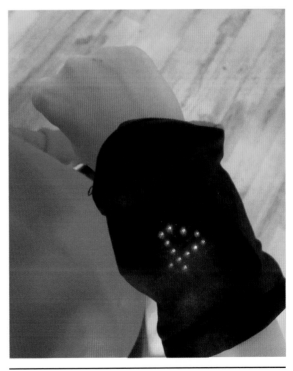

Figure 15.1   micro:bit step counter.

There are some great holders out there that people have made. I really like this one by Vlastimil Hovan. It keeps the micro:bit and battery together in a really sturdy case.

With your adult, watch a YouTube video of the holder at https://youtu.be/3mRmpCT9sCU, and look up Vlastimil on Twitter (https://twitter.com/Vlastimil_Hovan).

Code

Record the Steps

1. Create a variable called *steps*.
2. Change *steps* by 1 when the micro:bit shakes.
3. Display *steps*.

When we get more than nine steps, the numbers start scrolling on the micro:bit, and it's difficult to read them. Let's create a graph.

Create a Graph

The micro:bit lets you plot the LEDs using numbers. Each light has an address of two numbers X and Y. The grid on the micro:bit is laid out like this:

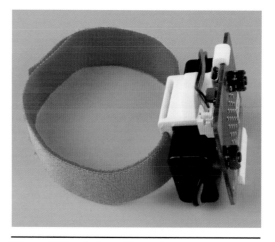

Figure 15.2   Holder with a Velcro strap.

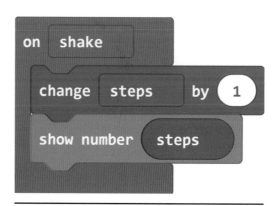

Figure 15.3   Step code.

0, 0	1, 0	2, 0	3, 0	4, 0
0, 1	1, 1	2, 1	3, 1	4, 1
0, 2	1, 2	2, 2	3, 2	4, 2
0, 3	1, 3	2, 3	3, 3	4, 3
0, 4	1, 4	2, 4	3, 4	4, 4

If we light up these coordinates using the block *plot x y* from the **LED** menu, like so:

X	Y
1	1
3	1
0	3
4	3
1	4
2	4
3	4

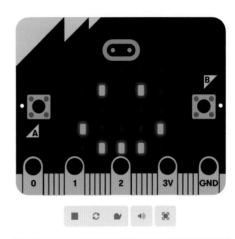

Figure 15.4   Smile.

we'll have a smiley face!

Let's graph out steps. When X gets to five steps, we need to increase Y by 1 and reset X to 0 to get to the next line. Here's the first line being filled in:

$X = 0, Y = 0$	$X = 1, Y = 0$	$X = 2, Y = 0$	$X = 3, Y = 0$	$X = 4, Y = 0$

When $X = 5$, we want to increase Y and put X back to 0 to start drawing our second line:

$X = 0, Y = 1$	$X = 1, Y = 1$	$X = 2, Y = 1$	$X = 3, Y = 1$	$X = 4, Y = 1$

And our third, fourth, and fifth lines . . .

$X = 0, Y = 2$	$X = 1, Y = 2$	$X = 2, Y = 2$	$X = 3, Y = 2$	$X = 4, Y = 2$
$X = 0, Y = 3$	$X = 1, Y = 3$	$X = 2, Y = 3$	$X = 3, Y = 3$	$X = 4, Y = 3$
$X = 0, Y = 4$	$X = 1, Y = 4$	$X = 2, Y = 4$	$X = 3, Y = 4$	$X = 4, Y = 4$

Let's create a new algorithm to help us figure this out:

```
On move
    Change steps by 1
    Change X by 1
    If X = 5
    Change Y by 1
    Set X to 0
Plot X and Y
```

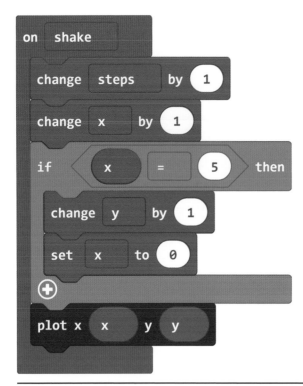

Figure 15.5   Plotting steps.

Debug

Once you start plotting the steps using *X* and *Y*, do you need the ***steps*** variable anymore? Did you test it in the simulator before downloading it to the micro:bit? Did you? Tell the truth! If you did, you would have spotted this bug:

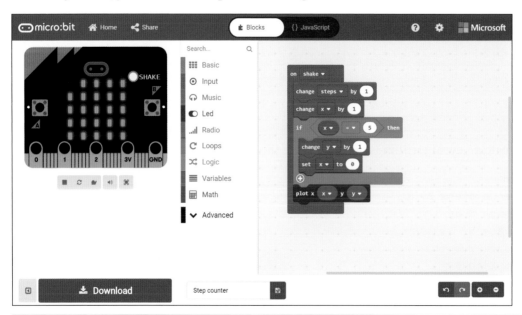

Figure 15.6   Spot the bug.

The first square isn't filled in! When you create a variable in MakeCode, it is set up as 0. When we first start moving, *X* increases to 1, and then it's plotted. We need *X* to start plotting at 0. Let's start *X* off as −1.

Expert Level

The grid fills up at 25 steps. That's not very far! How would you graph 1,000 steps?

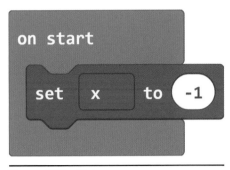

Figure 15.7   Debug.

Circuit Playground Express

Build

We want to attach the Circuit Playground Express to our wrist in a comfortable way.

Alternatively, the MakeCode team from Microsoft has a great tutorial on how to create a Circuit Playground Express wrist holder using duct tape and some Velcro. Follow the instructions at https://makecode.adafruit.com/projects/watch-timer.

Code

1. Create a variable called *steps*.
2. When the Circuit Playground Express shakes, change *steps* by 1.
3. Drag out *graph steps* from the **Lights** menu, and add the variable *steps*.

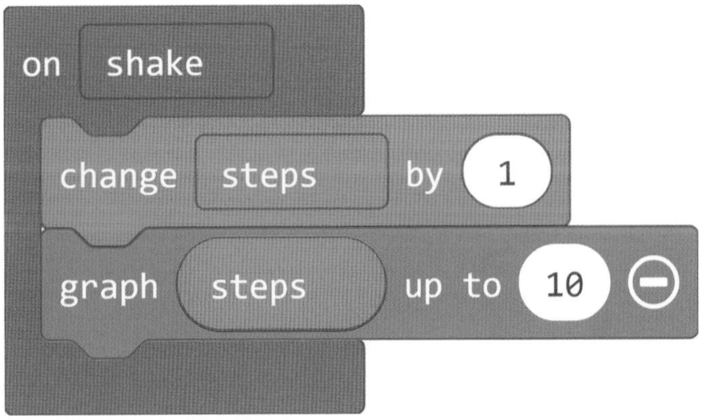

Debug

Do I … do I not have any bugs? OMG, that's a first!

Figure 15.8   Step code.

Expert Level

Graph up to a 100 steps. When you're halfway there, play a tune!

Bike Indicator

Cycling is great for the environment and for your own health and fitness. No smoke comes out of your bike when you cycle, and it doesn't use up resources such as gasoline, diesel fuel, or electricity to run it. All it takes is pedal power! To make cycling more fun and appealing, here's a fun indicator for your bike.

> Do not rely on this indicator as the only indicator for your bike. Use the correct hand signals, follow the road safety rules, use proper lights, and always wear a helmet when cycling. Be safe!

Cyclists are really hard to photograph! There are three different options for this mission:

1. A device as an indicator on the front of your bike
2. A device as an indicator on the front of your bike with arcade buttons
3. Two devices, one as an indicator on the back and one as buttons on the front.

You need to decide **NumberOfLoops** and **PauseLength** in the following algorithm. Have a look at indicators on cars and motorbikes where you live (safely from the sidewalk!). How fast do they flash? How long do they stay on? Research!

Figure 16.1   Just a blur!

Algorithm

```
On button A Press/Pin0 press/receive number 1
       Repeat NumberOfLoops times
              Show a left arrow
              Pause PauseLength
              Clear the screen
              Pause PauseLength
On button B Press/Pin 1 press/receive number 2
       Repeat NumberOfLoops times
              Show a right arrow
              Pause PauseLength
              Clear the screen
              Pause PauseLength
```

Build

Front Bike Indicator

Mount your device on the front of your bike. I used string through the pin holes to tie the indicator and battery on. Be careful here: if you attach the device to a metal bar, you might inadvertently connect ground to power—which is a disaster! Remember what I've been telling you all along? Ground to ground, power to power. We don't ever want to connect ground and power together. I put some electrical tape on the bike just beneath the device to make sure the bar wasn't conducting.

Arcade Button Indicator

For this build, you need the following:

- Four crocodile clips
- Two arcade buttons
- A box to hold everything in place

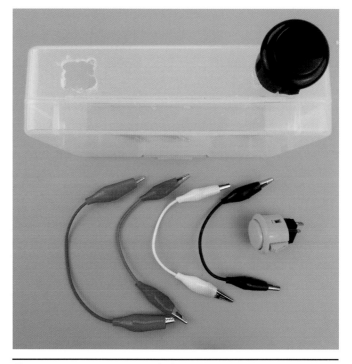

Figure 16.2   Equipment needed.

I used a plastic box to hold my device, crocodile clips, and arcade buttons in place. Plastic was really hard to work with because I had to use a drill to get the holes in the box to fit the buttons in. Drilling a plastic box is difficult because:

- Bits of hot plastic flew everywhere.
- The box was slippery to hold.
- It didn't look very pretty when done.

I chose clear plastic so that I could show you everything attached! You're probably better off using a cardboard box and sticking the device to the front. I used string and tape to hold everything in place. Make sure that the box is held on the handlebars securely before you set off.

1. Add the buttons to the top of the box.
2. Add your device to the front of the box, maybe using string to hold it in place. Cut a hole below or around the device for the crocodile clips to attach.
3. Connect the arcade buttons to the device. One leg of the arcade button attaches to GND. The other leg attaches to a data pin: 0, 1 or 2—it doesn't matter which leg.
4. Don't forget to make room for the battery.
5. Finally, make some holes in the box to attach it to the bike using string.

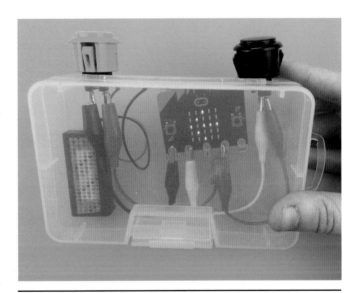

Figure 16.3 Device in a box.

micro:bit

Code

Indicator

Let's create the indicators on the micro:bit using its own buttons first.

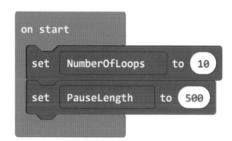

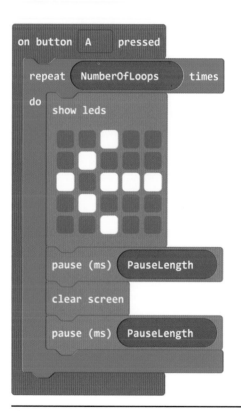
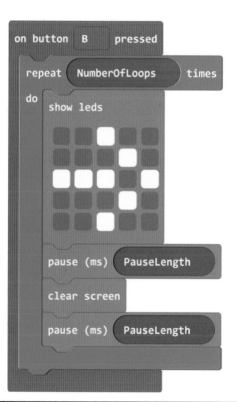

Figure 16.4   Using the micro:bit buttons.

Arcade Button Indicator

Here's the code for one button attached to pin 0. Create another *if* block below it for pin 1.

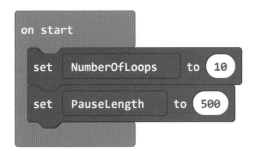

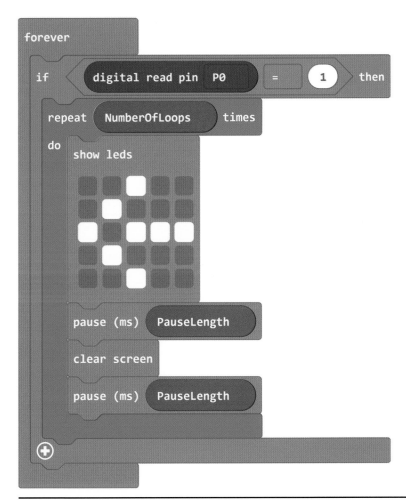

Figure 16.5 Arcade button code.

I really like the arcade button build. It makes it easier to press the buttons while you're cycling, and it looks really cool.

Back Indicator

Combine the buttons or arcade buttons with a second micro:bit and the radio blocks to create a back indicator.

1. On both micro:bits, set up the radio.
2. On the back micro:bit, display the arrows depending on what number you receive. Here's the code just for the left arrow. Can you work out the right arrow code?

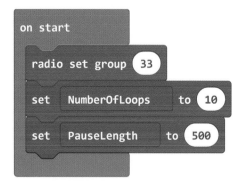

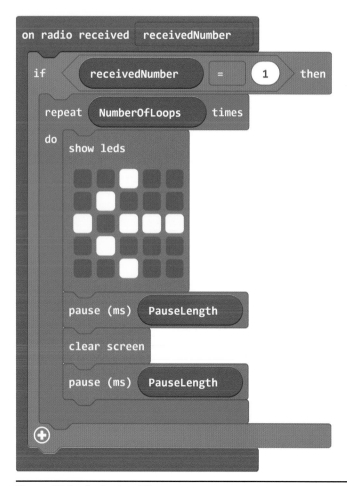

Figure 16.6 Back micro:bit.

3. Change the code on the front micro:bit.

This is the code for the micro:bit at the front to send the numbers when the micro:bit buttons are pressed:

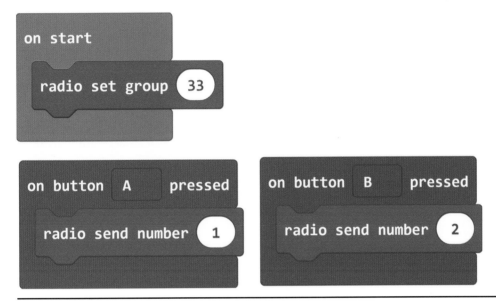

Figure 16.7   Front micro:bit code.

Here's the code for the micro:bit attached to the arcade buttons. I've kept the button presses in for testing.

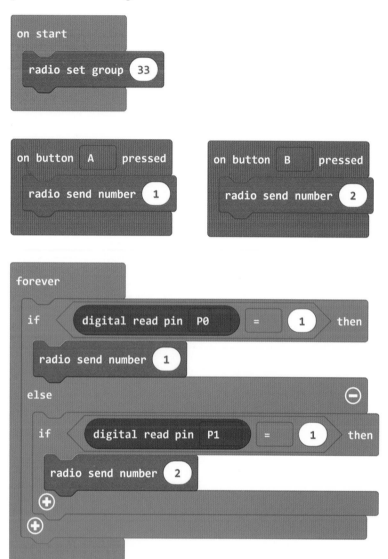

Figure 16.8   Front micro:bit arcade buttons.

Debug

For the mission with two micro:bits, I added the arrow to the front micro:bit just so I could tell which button I pressed! The first box that I had for this mission wasn't deep enough. With two crocodile clips on the 3-V pin, one of the crocodile clips sticks up, which stopped the lid from closing. I had to start again with a different box.

Expert Level

This will *really* impress your friends. Do you have the light strip from Missions 4 and 5? You know what to do.

Circuit Playground Express

I think the best part of creating an indicator is having the indicator on the back of your bike. With no line of site, the Circuit Playground Express is not going to be able to act as a back indicator. It makes a great front indicator, though. Replicate the earlier code to flash the LEDs left and right, just like the direction sign in Mission 2.

Figure 16.9   Do you have the light strip?

Moisture Sensor for Your Plants

Plants provide oxygen, they look good, and they smell great. Plants are cool. To keep my plants alive, I'm going to create a sensor that checks whether the soil inside my potted plants is moist.

When soil is wet, it conducts electricity! So we can use the micro:bit and the Circuit Playground Express to keep an eye on the moistness level of our potted plant soil. Try to say this three times: "potted plant soil, potted plant soil, potted plant soil."

Instead of using the ***digital read pin***, we're going to use ***analogue read***. *Digital read* returns either *1* or *0*, *true* or *false*. I'd like to know how wet/dry my soil is. I want a range of numbers. *Analogue read* returns a range of numbers. Both the micro:bit and the Circuit Playground Express have pins that can read analogue data.

Algorithm

We have two algorithms, one for getting our range and the other for determining the moisture content. We do this on dry, then wet soil.

```
Forever
        Display the analogue data on the pin connected to the pot
```

Once we have our range, we set up the pot forever! When the soil is dry, I want my alarm to go off until I water the plant, so a *while* loop is better here.

```
Forever
        While analogue read pin is dry
                Sound the alarm
        Pause for 15 minutes
```

Build

For this mission, you will need:

- A plant in a pot
- Two metal nails or screws
- Two crocodile clips

I found it easier to get a dry plant and measure it, and then water it and measure it again. It takes ages for a plant to dry out!

1. Stick the nails in the soil on opposite sides of the plant.
2. Attach a crocodile clip to each nail.

Figure 17.1 Potted plant ready to go.

micro:bit

Build

1. Attach one crocodile clip to GND on the micro:bit.
2. Attach the other crocodile clip to pin 1 on the micro:bit.
3. Attach a speaker to GND and pin 0 on the micro:bit.

Code

Let's get our range first. The *analogue read pin P1* block is in the **Pins** menu under the **Advanced** menu.

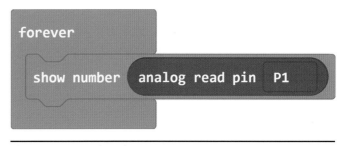

Figure 17.2 Finding the range.

I used a cactus that hadn't been watered in months as my dry plant. When I tested the plant, the number was around 170. I then used a plant that had just been watered as my wet plant. I got 1 for that number.

Here's my code. I set my "is dry" condition to be more than 50, because a cactus is quite an extreme example of a dry plant! If the plant is not dry, I don't check it again for 15 minutes. I don't think plants dry out that quickly, so you could make this number longer.

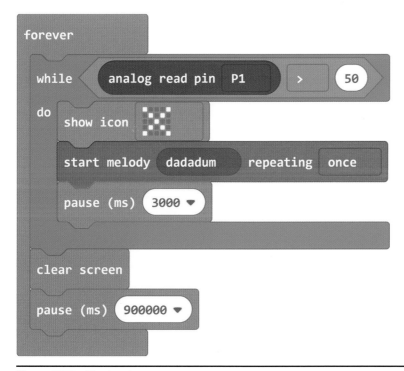

Figure 17.3 Plant code.

Debug

The whole point of using analogue was to know how wet/dry my plant was. I added the **read pin** to *A button pressed* and wrote the number 50 on my potted plant. In that way, I can check how close to dry my plant is.

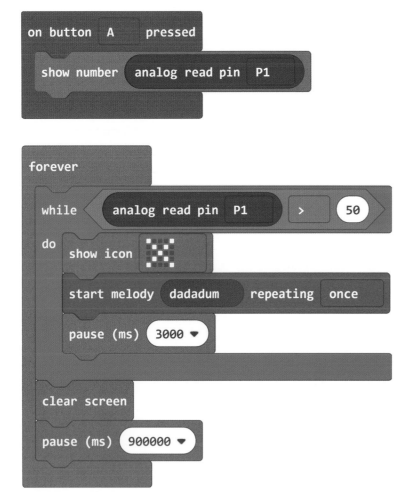

Figure 17.4 Debug.

Expert Level

You could use the micro:bit as a graph to show how dry the plant is. Use the plot code from Mission 15.

Circuit Playground Express

Build

1. Attach one crocodile clip to GND on the Circuit Playground Express.
2. Attach the other crocodile clip to pin A2 on the Circuit Playground Express.
3. Attach a speaker to GND and pin A0 on the Circuit Playground Express.

Code

With no screen from which to read numbers, we're going to have to run this code several times on the Circuit Playground Express. The analogue range is 0 to 1,023. I created the micro:bit code first and discovered that the dry plant produces larger numbers, so I tested this on the dry plant first.

I started graphing up to 1,023.

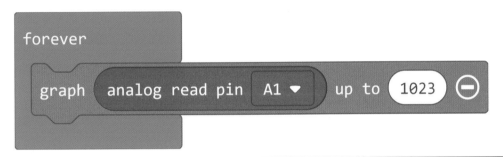

Figure 17.5 Graphing the plant.

This turned on only one light, so that's about 102? I changed the code to graph up to 100, and seven lights stayed on. I moved the nails to the damp plant and graphed up to 100—two lights stayed on.

Let's say then that 70 is dry and 20 is wet. If my plant is above 50, I would say it's dry.

Here's the code to check for dryness and sound the alarm. If the plant is wet, it won't check again until 15 minutes later. This code has the alarm constantly going off until you water the plant. What will yours do? Sound the alarm and then pause for a minute and sound it again?

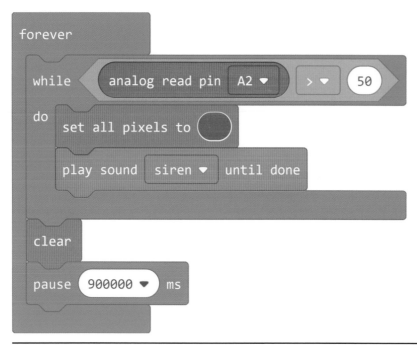

Figure 17.6 Plant code.

Debug

I wanted to see how close to dry my plant was, so I put the graph code into a button press. I cleared the light after 5 seconds, so that it doesn't stay on for up to 15 minutes.

Figure 17.7 Debugged.

Expert Level

How often does your plant need watering? Can you record how often you water the plant in a week?

Temperature Monitor

A lot of energy is created, spent, and wasted on heating our homes. Don't get me wrong, we need heat, and I love a nice cosy house. Sometimes, though, my house is too hot, and I'm wasting energy. Let's set up a temperature sensor to get an idea of what temperature your house actually is.

Instead of displaying the current temperature, let's record the temperature every hour and store it in an array. Then we can do some science with it!

ARRAY
An *array* is like a list. You can store numbers in an array and pull them out later. So instead of having variables **temperature1**, **temperature2**, and **temperature3**, we have an array called **temperatures** that stores all those numbers inside it. The great thing about an array in MakeCode, as opposed to other programming languages, is that you don't have to set the size when you start, and if you do, the array can grow in size later on in your code.

Algorithm

The number of hours you want to record for is *X*.

```
On start
    Repeat X times:
        Get the temperature
        Add the temperature to the array
        Pause for an hour
    Display a tick
On button A press
    Loop through the array
        Show the temperature
```

micro:bit

Build

Place the micro:bit somewhere inside your house, away from a radiator or open windows. Try not to hold the micro:bit in your hand either.

Figure 18.1 The micro:bit says it's too hot!

Code

1. Let's have a look at **Arrays** in MakeCode. They're under **Advanced** at the very bottom of the menu.
2. Drag out the block **set list to array of 1 2**, and place it under **on start**.

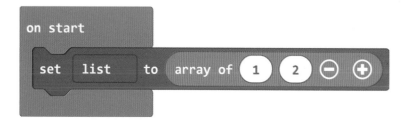

Figure 18.2 From the **Array** menu.

3. Press the minus sign (–) twice to get an empty array.
4. Let's give our array a better name. Click on **list**, and select **Rename variable**.
5. Type in "Temperatures," and click *OK*.

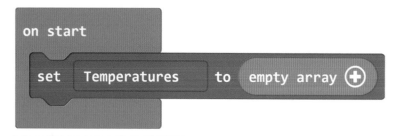

Figure 18.3 Set up an empty temperature array.

6. Now we have an empty array that we can fill up with numbers.
7. From **Array**, drag out the block **list add value to end**, and place it under **on start**.

8. Change *list* to *Temperatures*.

9. From the **Input** menu, drag out *temperature (C)*, and place it in the blank spot after *value*.

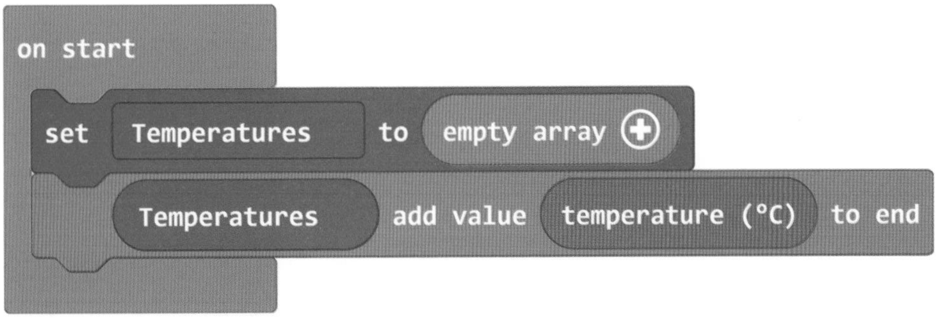

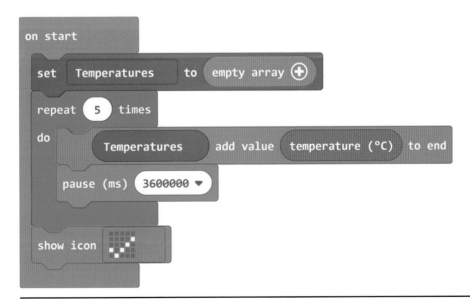

Figure 18.4 Add temperature to the array.

10. We're going to repeat this five times with an hour pause in between each recording. Put a repeat block around these blocks.

1,000 ms = 1 second

60,000 ms = 1 minute

3,600,000 ms = 1 hour

11. After the array is finished, we'll insert a tick so that I know when it's done. That's recording the temperature for 5 hours in the array. When it's done, we want to get the data out.

Figure 18.5 Record every hour.

12. Under **Loops**, there is a loop to go through an array *for element value of list*.
    a. *Value* is the number we've stored inside the array.
    b. *List* is the name of the array.
13. Put this block under *on button A pressed*.
14. Change *list* to *Temperatures*.
15. Add *show number* with the variable *value* inside this loop.

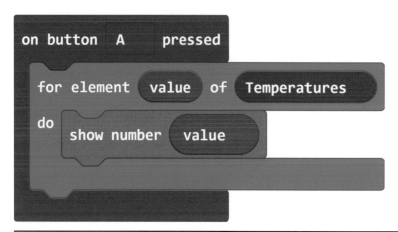

Figure 18.6   Looping through the array.

Once the micro:bit starts, it will record temperatures and store them. When you press A, you'll be able to read the temperatures recorded so far.

Debug

There are a couple of issues here. The temperature is in degrees Celsius. If you want degrees Fahrenheit, you're going to have to use some blocks from the **Math** menu. The math is (Temperature in Celsius × 9/5) + 32.

To test your code, instead of standing outside in the cold for an hour, you can:

- Change your pause to just 5 seconds.
- Hold the micro:bit tightly in your hand to increase the temperature.

Even when you have the temperature in the right scale, do you find that your temperature readings are far too high? You don't have a fever; the micro:bit's temperature readings are a bit higher than the actual room temperature. This is because it's recording the temperature of the micro:bit, and the micro:bit is an electrical device that produces heat.

You need to find the *offset*—the difference between the micro:bit temperature and the actual room temperature. You will need a real thermometer for this task.

1. Place the micro:bit and a real thermometer in the same location.
2. Get the micro:bit temperature.

3. Get the temperature that the thermometer shows.

4. Do the same every hour for a few hours.

5. Create a grid, and enter your numbers. Here's mine:

micro:bit	Thermometer
24	21
23	21
23	20
22	19

6. Add a third column, and work out the difference between the micro:bit's temperature readings and those of the thermometer.

micro:bit	Thermometer	Difference
24	21	3
23	21	2
23	20	3
22	19	3

On average, my difference was 3 degrees. Your code now needs to take this offset into account. Let's record the temperatures correctly. I added a **_math_** block when I added the temperature to the array.

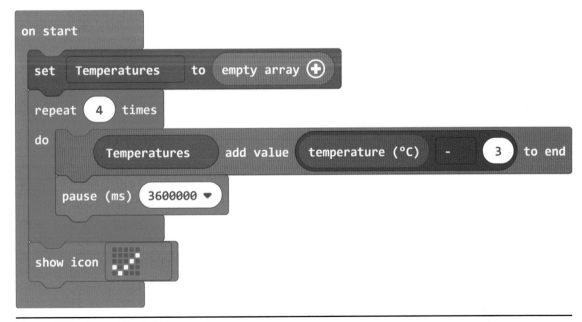

Figure 18.7 Record the right temperatures.

Expert Level

In Mission 19, you can create an alert if your temperature is too hot. And Mission 20 uses both a sensor on a window and this sensor.

I really like the grid of temperatures that we recorded earlier. Why not record your temperatures over 24 hours, fill out a grid, reset the micro:bit, and do it again? What does a week's worth of temperature data in your house look like? Can you create a line graph? Is your grandparents' house hotter than that of your parents? What about your school? Which room in your house is the warmest? There are lots of experiments you can run with this sensor. Try them out!

Circuit Playground Express

Build

Place the Circuit Playground Express in your house away from any heat sources such as a radiator or the dog.

Code

Let's have a look at arrays in MakeCode. They're under **Advanced** at the very bottom of the menu.

1. Drag out the block *set list to array of 1 2*, and place it under *on start*.

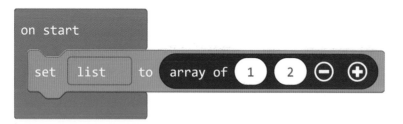

Figure 18.8   From the **Array** menu.

2. Press the minus sign (–) twice to get an empty array.
3. Let's give our array a better name. Click on *list*, and select *Rename variable*.
4. Type in "Temperatures," and click *OK*.

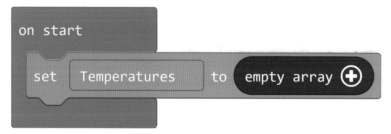

Figure 18.9   Set up an empty temperature array.

5. Now we have an empty array that we can fill up with numbers.
6. From **Array**, drag out the block *list add value to end*, and place it under *on start*.
7. Change *list* to *Temperatures*.
8. From the **Input** menu, drag out *temperature in C*, and place it in the blank spot after *value*.

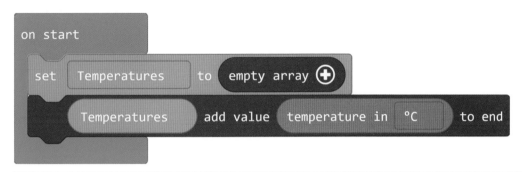

Figure 18.10   Add temperature to the array.

9. I'm going to repeat this four times with an hour pause in between each recording.
10. After the array is finished, I'll set all the lights to green so that I know when it's done. That's recording the temperature for four hours into the array. When it's done, we want to get the data out.

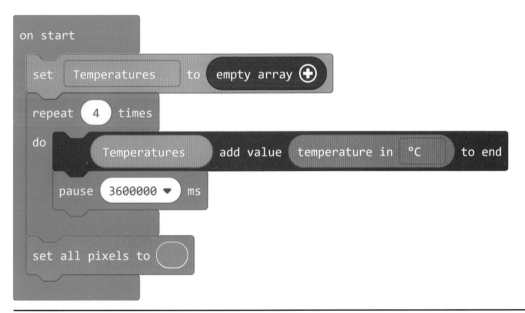

Figure 18.11   Record every hour.

11. Under **Loops**, there is a loop to go through an array: *for element value of list*.
   a. *Value* is the number we've stored inside the array.
   b. *List* is the name of the array.
12. Put this block under *on button A pressed*.
13. Change *list* to *Temperatures*.

We're going to display the temperatures on the ring of lights around the Circuit Playground Express using the *graph 0* block found under **Light** menu.

The temperature in my house is between 15 and 25 degrees Celsius. To graph this on the Circuit Playground Express, I need to graph *temperature - 15 out of 10*. This would give a light to every temperature between 16 and 25. Thus 16 degrees would be one light, 17 degrees would be two lights, and so on. I've put a pause in here so that we see the graph before it's overwritten with a different color.

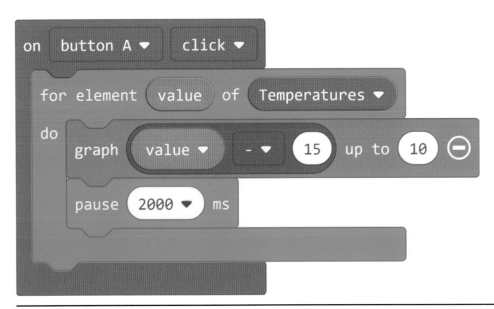

Figure 18.12   Looping through the array.

In the north of England, where I live, 25 degrees Celsius is about as hot as it gets! If your house is hotter or even cooler, change these numbers.

Debug

To test this code, I changed the pause between temperatures to be just 10 seconds. Then to heat up the Circuit Playground Express, I held it tightly in my hand. I saw the temperature change quickly using this technique.

Without holding the Circuit Playground Express in your hand, do you find that your temperature readings are far too high? You don't have a fever. The Circuit Playground Express temperature readings are a bit higher than the actual room temperature. This is because it's recording the temperature of the board, and the board is an electrical device that produces heat.

You need to find the *offset*—the difference between the Circuit Playground Express temperature and the real room temperature. You will need a real thermometer for this experiment.

1. Place the Circuit Playground Express and a real thermometer in the same location.
2. Get the Circuit Playground Express temperature.
3. Get the temperature that the thermometer shows.
4. Do the same in an hour's time for a few hours.
5. Create a grid, and enter your numbers. Here's mine:

Circuit Playground Express	Thermometer
24	21
23	21
23	20
22	19

6. Add a third column, and work out the difference between the Circuit Playground Express temperature readings and those of the thermometer.

Circuit Playground Express	Thermometer	Difference
24	21	3
23	21	2
23	20	3
22	19	3

On average, my difference was 3 degrees. Your code now needs to take this offset into account. Let's record the temperatures correctly. I added a *minus* block from the **Math** menu when I added the temperature to the array.

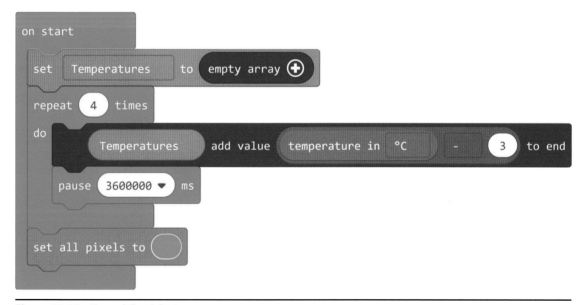

Figure 18.13   Record the right temperature.

Expert Level

Go to Mission 19 to create an alert if your temperature is too hot. Mission 20 uses a window sensor and this sensor together.

The grid of temperatures we created earlier is really interesting. I found out that my house starts getting warmer at 5 a.m., and yet I don't wake up until 7 a.m. Record your house temperatures over 24 hours, and fill in your own grid. Then do it again for the week. Find out the patterns of your house. Try recording different houses, such as the house of one of your friends or your grandparents, and compare the temperatures. Or just compare different rooms in your house. Is the kitchen the warmest room in the house? There are lots of experiments you can run with this sensor. Try them out!

Raspberry Pi

Build

Your Raspberry Pi doesn't have a temperature sensor, but it does have access to the internet. We're going to find out the outside temperature to help us get our inside temperature right.

Code

We're still going to follow most of our original algorithm. Get the temperature every hour, and store it in an array.

Figure 18.14   So cold!

Get Your Longitude and Latitude

To find the temperature outside your house, you need to know the longitude and latitude of your area. I find the best way to do this is to use Google Maps.

1. Go to http://maps.google.com.
2. Search for your home address, and then zoom in on your house.
3. Right-click on your house on the map.
4. Select "**What's here?**"
5. A little box pops up at the bottom of the screen with the longitude and latitude numbers.
6. Write these down.

Here's where I don't live! Buckingham Palace's longitude and latitude: 51.501401, −0.141895.

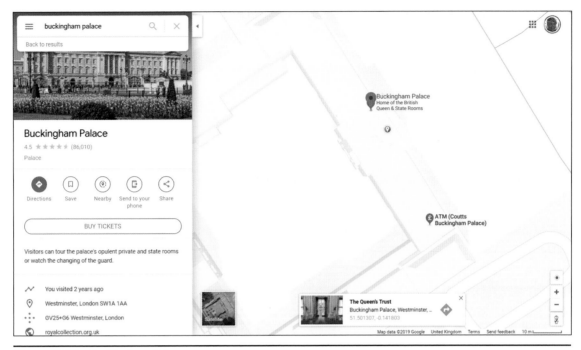

Figure 18.15 Buckingham Palace.

Set Up Your API

We've got our coordinates. Now we need to get the temperature. We're going to use an API.

API
API stands for "application programming interface." APIs are used when two pieces of code want to "talk" to each other. In this case, your temperature code wants to talk to a piece of code on the Dark Sky computers.

You will need an adult to sign up for an account with Dark Sky. You can get data from this API 1,000 times a day for free. You cannot go over 1,000 with a free account. Your adult will not be billed for it, so he or she does not have to enter any billing information. Here are the instructions:

1. Go to https://darksky.net/dev.
2. Select *Sign up*, and enter your email address and a password.
3. You will need to check your email for a link to confirm your account.
4. Once your account is confirmed, log in with your email address and password.
5. When you are logged in, it will show you your secret key.
6. Copy your secret key from the website.

7. On the same page, there's a link like this: https://api.darksky.net/forecast/YOURSECRETKEY/37.8267,-122.4233.
8. When you click it, it gives you all the weather data for Los Angeles.
9. Change the last part of the address to your longitude and latitude. For example, Buckingham Palace would be https://api.darksky.net/forecast/YOURSECRETKEY/51.501401,-0.141895.

There's a lot of information on this web page! Can you find the emperature? It's on the second line.


```
{"latitude":51.501401,"longitude":-0.141895,"timezone":"Europe/London","currently":
{"time":1556028564,"summary":"Overcast","icon":"cloudy","nearestStormDistance":100,"nearestStormBearing":177,"precipIntensity":0,"precipProbability":0,"temperature":68.01,"ap
parentTemperature":68.01,"dewPoint":44.23,"humidity":0.42,"pressure":996.35,"windSpeed":10.21,"windGust":15.06,"windBearing":78,"cloudCover":0.94,"uvIndex":2,"visibility":6.3
6,"ozone":371.7},"minutely":{"summary":"Overcast for the hour.","icon":"cloudy","data":[{"time":1556028540,"precipIntensity":0,"precipProbability":0},
{"time":1556028600,"precipIntensity":0,"precipProbability":0},{"time":1556028660,"precipIntensity":0,"precipProbability":0},
{"time":1556028720,"precipIntensity":0,"precipProbability":0},{"time":1556028780,"precipIntensity":0,"precipProbability":0},
{"time":1556028840,"precipIntensity":0,"precipProbability":0},{"time":1556028900,"precipIntensity":0,"precipProbability":0},
{"time":1556028960,"precipIntensity":0,"precipProbability":0},{"time":1556029020,"precipIntensity":0,"precipProbability":0},
{"time":1556029080,"precipIntensity":0,"precipProbability":0},{"time":1556029140,"precipIntensity":0,"precipProbability":0},
{"time":1556029200,"precipIntensity":0,"precipProbability":0},{"time":1556029260,"precipIntensity":0,"precipProbability":0},
{"time":1556029320,"precipIntensity":0,"precipProbability":0},{"time":1556029380,"precipIntensity":0,"precipProbability":0},
{"time":1556029440,"precipIntensity":0,"precipProbability":0},{"time":1556029500,"precipIntensity":0,"precipProbability":0},
{"time":1556029560,"precipIntensity":0,"precipProbability":0},{"time":1556029620,"precipIntensity":0,"precipProbability":0},
{"time":1556029680,"precipIntensity":0,"precipProbability":0},{"time":1556029740,"precipIntensity":0,"precipProbability":0},
{"time":1556029800,"precipIntensity":0,"precipProbability":0},{"time":1556029860,"precipIntensity":0,"precipProbability":0},
{"time":1556029920,"precipIntensity":0,"precipProbability":0},{"time":1556029980,"precipIntensity":0,"precipProbability":0},
{"time":1556030040,"precipIntensity":0,"precipProbability":0},{"time":1556030100,"precipIntensity":0,"precipProbability":0},
{"time":1556030160,"precipIntensity":0,"precipProbability":0},{"time":1556030220,"precipIntensity":0,"precipProbability":0},
{"time":1556030280,"precipIntensity":0,"precipProbability":0},{"time":1556030340,"precipIntensity":0,"precipProbability":0},
{"time":1556030400,"precipIntensity":0,"precipProbability":0},{"time":1556030460,"precipIntensity":0,"precipProbability":0},
{"time":1556030520,"precipIntensity":0,"precipProbability":0},{"time":1556030580,"precipIntensity":0,"precipProbability":0},
{"time":1556030640,"precipIntensity":0,"precipProbability":0},{"time":1556030700,"precipIntensity":0,"precipProbability":0},
{"time":1556030760,"precipIntensity":0,"precipProbability":0},{"time":1556030820,"precipIntensity":0,"precipProbability":0},
{"time":1556030880,"precipIntensity":0,"precipProbability":0},{"time":1556030940,"precipIntensity":0,"precipProbability":0},
{"time":1556031000,"precipIntensity":0,"precipProbability":0},{"time":1556031060,"precipIntensity":0,"precipProbability":0},
{"time":1556031120,"precipIntensity":0,"precipProbability":0},{"time":1556031180,"precipIntensity":0,"precipProbability":0},
{"time":1556031240,"precipIntensity":0,"precipProbability":0},{"time":1556031300,"precipIntensity":0,"precipProbability":0},
{"time":1556031360,"precipIntensity":0,"precipProbability":0},{"time":1556031420,"precipIntensity":0,"precipProbability":0},
{"time":1556031480,"precipIntensity":0,"precipProbability":0},{"time":1556031540,"precipIntensity":0,"precipProbability":0},
{"time":1556031600,"precipIntensity":0,"precipProbability":0},{"time":1556031660,"precipIntensity":0,"precipProbability":0},
{"time":1556031720,"precipIntensity":0,"precipProbability":0},{"time":1556031780,"precipIntensity":0,"precipProbability":0},
{"time":1556031840,"precipIntensity":0,"precipProbability":0},{"time":1556031900,"precipIntensity":0,"precipProbability":0},
{"time":1556031960,"precipIntensity":0,"precipProbability":0},{"time":1556032020,"precipIntensity":0,"precipProbability":0},
{"time":1556032080,"precipIntensity":0,"precipProbability":0},{"time":1556032140,"precipIntensity":0,"precipProbability":0}]},"hourly":{"summary":"Mostly cloudy until
tomorrow afternoon.","icon":"partly-cloudy-night","data":[{"time":1556028000,"summary":"Mostly Cloudy","icon":"partly-cloudy-
day","precipIntensity":0,"precipProbability":0,"temperature":68.08,"apparentTemperature":68.08,"dewPoint":44.4,"humidity":0.42,"pressure":996.4,"windSpeed":10.18,"windGust":1
4.96,"windBearing":78,"cloudCover":0.93,"uvIndex":3,"visibility":6.22,"ozone":371.66},
{"time":1556031600,"summary":"Overcast","icon":"cloudy","precipIntensity":0,"precipProbability":0,"temperature":67.67,"apparentTemperature":67.67,"dewPoint":43.31,"humidity":
0.41,"pressure":996.08,"windSpeed":10.36,"windGust":15.61,"windBearing":80,"cloudCover":1,"uvIndex":2,"visibility":7.13,"ozone":371.96},
{"time":1556035200,"summary":"Overcast","icon":"cloudy","precipIntensity":0,"precipProbability":0,"temperature":69.09,"apparentTemperature":69.09,"dewPoint":42.77,"humidity":
0.39,"pressure":995.79,"windSpeed":10.25,"windGust":15.84,"windBearing":84,"cloudCover":1,"uvIndex":1,"visibility":10,"ozone":372.33},
{"time":1556038800,"summary":"Overcast","icon":"cloudy","precipIntensity":0,"precipProbability":0,"temperature":69.4,"apparentTemperature":69.4,"dewPoint":42.87,"humidity":0.
38,"pressure":995.53,"windSpeed":9.71,"windGust":15.8,"windBearing":86,"cloudCover":1,"uvIndex":1,"visibility":10,"ozone":372.69},
{"time":1556042400,"summary":"Overcast","icon":"cloudy","precipIntensity":0,"precipProbability":0,"temperature":67.52,"apparentTemperature":67.52,"dewPoint":43.25,"humidity":
0.41,"pressure":995.35,"windSpeed":8.93,"windGust":15.61,"windBearing":86,"cloudCover":0.98,"uvIndex":0,"visibility":10,"ozone":373.16},
{"time":1556046000,"summary":"Overcast","icon":"cloudy","precipIntensity":0,"precipProbability":0,"temperature":65.58,"apparentTemperature":65.58,"dewPoint":43.42,"humidity":
```


Figure 18.16 Found it!

There are other temperature readings on this page, but this is the current temperature in degrees Fahrenheit. If you want Celsius, you need to add an extra bit to the end of the URL: ?units=si.

Code the Raspberry Pi

We're going to code some Python to grab this number from the web for our mission.

Here's the code; for now, it's just printing out the current temperature. A lot of the code is getting all that text from the web page and pulling out that one number into a variable ***currentTemp***.

```
1  from urllib.request import urlopen
2  import json
3
4
5  #fill in these fields with your data
6  apikey="YOURSECRETKEYGOESHERE"
7  lati="51.501401"
8  longi="-0.141895"
9
10
11 #here's the web address to get the data from the API
12 url="https://api.forecast.io/forecast/"+apikey+"/"+lati+","+longi+"?units=si"
13
14 #let's try getting the temperature.
15 try:
16     meteo=urlopen(url).read()
17     meteo = meteo.decode('utf-8')
18     weather = json.loads(meteo)
19     currentTemp = weather['currently']['temperature']
20
21     print ("The temperature is ", currentTemp)
22
23 except IOError:
24     print ("Your Internet connection is not working!")
```

Figure 18.17   Code.

1. To get the code working, change the following variables to your data:

   a. *apikey*

   b. *lati*

   c. *longi*

2. Run the code to see if it returns the temperature that we saw on the website.

> **TRY AND EXCEPT**
>
> Did you notice the try and except code? This helps our code run even when there is an error. The error could be that we have no internet connection. This code tries to run the internet code, but if it can't, it prints out the error and continues in the code. It doesn't crash. It's a neat way of catching errors without crashing your whole program.

3. Now that we have the number, let's store it in a list. At first, let's do this every 5 seconds, and let's print the list to make sure that it's all working. Here's what I did:

   a. Added the *time* library at the top of the code

   b. Created an empty list called *temperatures*

   c. Created a *for* loop and appended the *currentTemp* variable to the *temperatures* list (I added *0* if the internet connection was down that hour.)

   d. Paused for 5 seconds

   e. Printed the list

```
1  from urllib.request import urlopen
2  import json
3  import time
4
5  apikey="YOURSECRETKEYGOESHERE"
6  lati="51.501401"
7  longi="-0.141895"
8
9
10 temperatures = []
11
12
13 #get the data from the api website
14 url="https://api.forecast.io/forecast/"+apikey+"/"+lati+","+longi+"?units=si"
15
16 #get the weather 5 times
17 for i in range(5):
18     try:
19             meteo=urlopen(url).read()
20             meteo = meteo.decode('utf-8')
21             weather = json.loads(meteo)
22             currentTemp = weather['currently']['temperature']
23
24             print ("The temperature is ", currentTemp)
25             temperatures.append(currentTemp)
26
27     except IOError:
28             print ("Your Internet connection is not working!")
29             temperatures.append(0)
30
31     #every 5 seconds
32     time.sleep(5)
33
34 print(temperatures)
```

Figure 18.18 Code.

4. Run the code to see what happens. After 25 seconds, you should have a list full of temperatures.

5. Here's a nicer way of printing lists.

6. The temperature is printed as we record it, and then, when all the temperatures are recorded, the entire list is printed again.

```
1  #print all the temperatures
2  for t in temperatures:
3          print(t)
```

Figure 18.19 Code.

```
                    Python 3.5.3 Shell              _ □ ×
File  Edit  Shell  Debug  Options  Window  Help
Python 3.5.3 (default, Sep 27 2018, 17:25:39)
[GCC 6.3.0 20170516] on linux
Type "copyright", "credits" or "license()" for more information.
>>>
========================= RESTART: /home/pi/C.py =========================
The temperature is  19.92
The temperature is  19.92
The temperature is  19.92
The temperature is  19.92
The temperature is  19.92
19.92
19.92
19.92
19.92
19.92
>>> |
```

Figure 18.20 Temperature readouts.

7. If you're happy with the code, change:
   a. The range to however long you want to record the temperature for
   b. The amount of time between each recording
8. For example, to record every hour for 24 hours, change:
   a. 5 to 24 in *for in range(5)*
   b. 5 to 3,600 in *time.sleep(5)*

Debug

I made lots of mistakes with the URL:

https://api.darksky.net/forecast/YOURSECRETKEY/51.501401,-0.141895

I left out the full stop (the period) and the comma. Without them, it just doesn't look like a URL that will work!

Expert Level

Have a look at the other data on the API page. What else could you pull out? For example, *precipProbability* is how likely it is going to rain. If this is over a certain number, you could set off an alarm that it's about to rain. In the code, instead of *currentTemp = weather['currently']['temperature']*, you could get other data such as *weather['currently'] ['precipProbability']* or *weather['currently']['windGust']* and even a summary: *weather['currently']['summary']*. The website https://darksky.net/dev/docs#forecast-request gives you more information on the data.

Temperature Alarm

Ha-ha, an empty chapter! What's too hot for you? What's too cold? You could record the temperature of your house over 24 hours and work out the average temperature. If the temperature goes 5 degrees over or under the average, then you could use those numbers as hot and cold.

Create an alarm for when your house is too hot or too cold. Send a different message or get a different alert depending on whether it's too hot or too cold. When the temperature is too hot, show a happy face on the micro:bit or red lights on the Circuit Playground Express and have some friends over for some juice. When it's too cold, show a sad face on the micro:bit or blue lights on the Circuit Playground Express, and put on a woolly hat.

Figure 19.1   Too cold!

Use the alarm code you've gathered from Missions 12, 13, and 14 to help you complete this mission. If you've done those missions already, then you should have a good idea of how to complete this mission. You've got this!

Window Alarm

When windows are open, heat can escape. But fresh air is nice, even in the cold. We don't want alarms going off just because a window is open. We just need to make sure that the window isn't open for too long.

We can use the temperature sensor from Mission 17 and the door sensor from Mission 8 together to accomplish this mission.

Algorithm

We want the alarm to go off if the window is open and the temperature in the house has gone down since the last two readings.

```
Every hour
      Get temperature and store in array
Forever
      When window is opened then
            Start timer
Forever
      If window is open AND timer > 15 minutes then
            Get last two temperature from the array
            If currentTemp < lastTemp AND currentTemp < secondLastTemp then
                  Signal alarm
```

micro:bit

Build

Add the tinfoil and crocodile clips to the edge of your window. Follow the instructions from Mission 8 on how to do this.

Figure 20.1   Window set up with tinfoil.

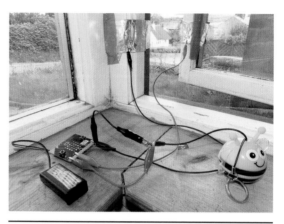

Figure 20.2   The micro:bit attached to the window with a speaker.

Code

1. Let's grab the code from Mission 18 to get the temperature and add it to an array.

```
on start
    set  Temperatures  to  empty array ⊕
    repeat  4  times
    do
            Temperatures  add value  temperature (°C)  -  3  to end
        pause (ms)  3600000 ▼
```

Figure 20.3   Code from Mission 18.

2. We'll need to change this to record the temperature forever. Take out the code inside *repeat 4 times* and put it inside a *forever* block.

3. Now let's check whether the window is open or closed. We're using the block *running time (ms)* from the menu **Inputs** then **more**.

4. I've added the letters *O* and *C* to the code to let me know the position of the window.

5. *windowOpen* is *true* when the window is open and *false* when the window is closed. The block *on pin P1 released* is under the menu **Inputs** then **more**.

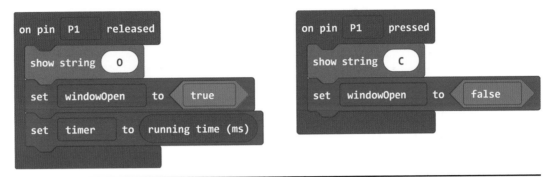

Figure 20.4 Setting the window state.

6. Finally, if the window is open and our timer is over 15 minutes, get the last two temperatures from the array. I created variables *Last Temp* and *Second Last Temp* to store these values.

In an array, variables have an address. This address tells us where a variable is. Only one variable can exist in one address. So it's a unique way of finding a variable. We start counting addresses from 0. Here's an array with addresses:

Address	Variable
0	17
1	18
2	14

The array size is 3: there are three variables in the array. The last variable in this array is at position 2, which is the size of the array minus 1. The second-to-last variable in this array is at position 1, which is the size of the array minus 2.

7. I created another variable *ArraySize* to record the size of the array.

8. If the window is open *and* the current time is greater than the timer + 15 minutes, get the last and second-to-last temperatures.

9. Instead of checking *if windowOpen = true*, we can just use *if windowOpen*.

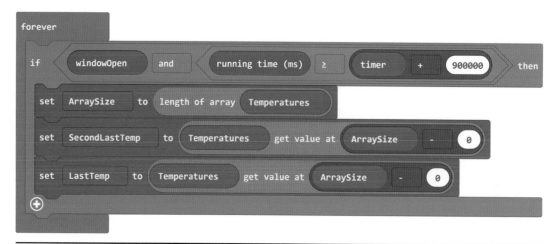

Figure 20.5 Getting the last two temperatures.

10. Check these temperatures against the current temperature.
11. Signal the alarm.

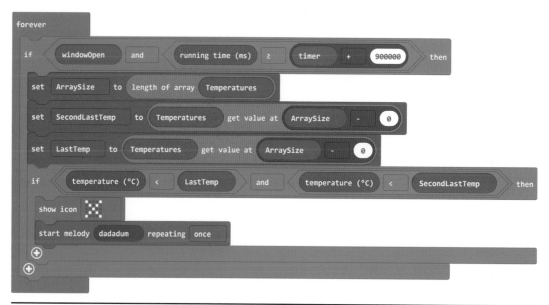

Figure 20.6 If it's cold, sound the alarm!

Debug

We need to reset the timer back to the current running time, or else the alarm will go off constantly. Also, remember how we had to subtract an offset to get the *real* temperature? We don't need to do this here, because we are just looking for the *change* in temperature. Here's all the code together:

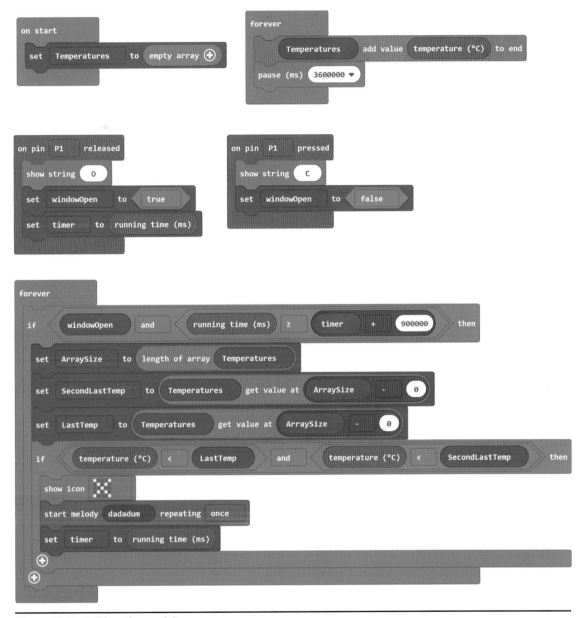

Figure 20.7 Full bug-free code?

If your alarm never goes off, you could change the code. You could check only the last temperature rather than both the last and second to last. You could check after the window is open for an hour. My alarm only went off when it was getting dark and cooler outside, which is a good time to close your window.

There's a second bug here that I'm not going to fix. If the window is open for 15 minutes and there is only one temperature recorded, what happens? You tell me!

In some programming languages, if you try to access an address that doesn't exist in the array, the program crashes. What happens in MakeCode? Does it affect your program? If it doesn't affect your program, should you fix it?

Expert Level

Oh wow, if you've created all that code and got it working with the window, I don't think there is an expert level for you. You are an expert!

Circuit Playground Express

Build

Add the tinfoil and crocodile clips to the edge of your window and the frame. Follow the instructions from Mission 8 on how to connect the window and the speaker to the Circuit Playground Express.

Code

We're basing this code on the code from Mission 18 to get the temperature and add it to an array. But there are a lot of changes too.

1. We need to set pin A2 to pull up.
2. I'm recording the temperature of the Circuit Playground Express, not the room. Remember how they're different? Later on in the code, we will check the temperature twice. Instead of always subtracting the offset, I'm just checking the Circuit Playground Express temperature.
3. Instead of just grabbing four temperature recordings, I'm going to run this code forever.
4. In Mission 18, we had code under **button A click**. We don't need that code here.

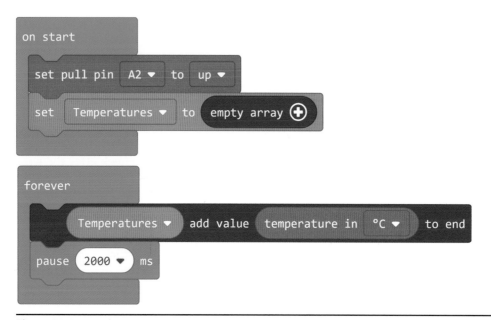

Figure 20.8   Get the temperature.

5. I'm getting the temperature every 2 seconds here: that's so that I can test my code quicker instead of waiting for an hour.

6. Now let's check whether the window is open or closed. We're using the block *timer 1 reset* from the **Advanced** menu and then **Control**.

7. I've added the colors red for open and green for closed to let me know what the Circuit Playground Express thinks the window has done.

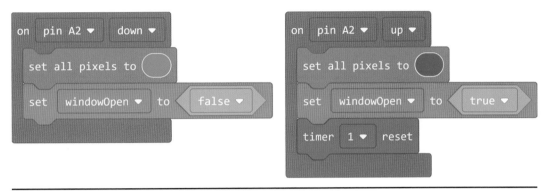

Figure 20.9 Opening and closing the window.

8. Finally, if our timer is over 15 minutes, get the last two temperatures from the array. I created variables *LastTemp* and *SecondLastTemp* to store these. Check out the box on page 239 of this mission to learn about arrays and addresses.

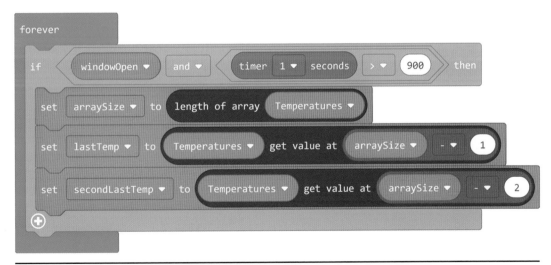

Figure 20.10 Getting the last two temperatures.

9. Check these temperatures against the current temperature.

10. Signal the alarm.

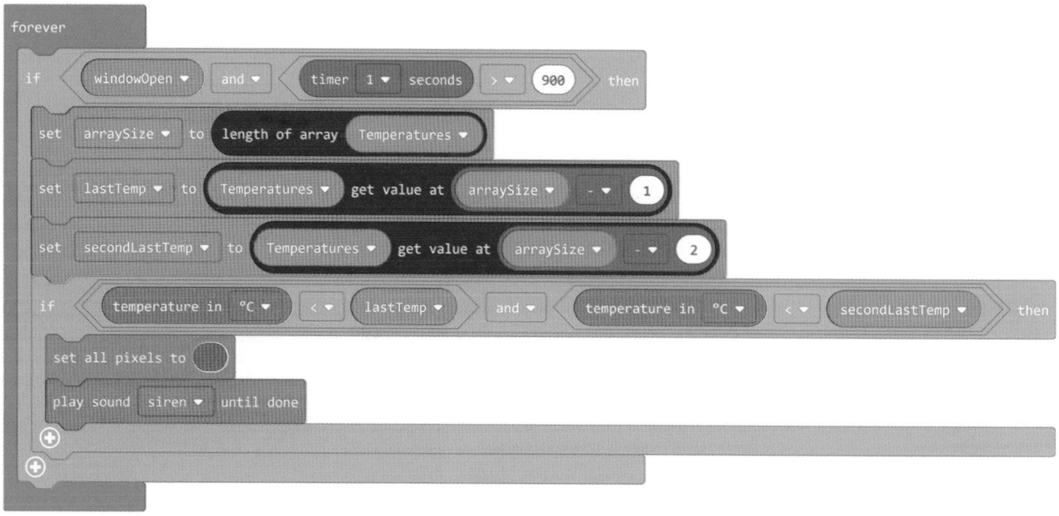

Figure 20.11   If it's cold, sound the alarm!

Debug

We need to reset the timer, or else the alarm will go off constantly.

When testing the code, you might want to set the number 900 seconds (15 minutes) to something like 10 seconds.

Expert Level

Seriously, look at that code in Figure 20.12. There is nothing more I can teach you!

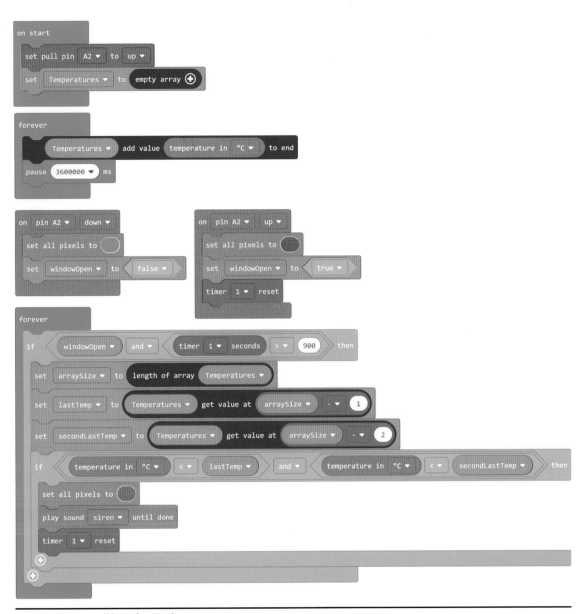

Figure 20.12 Full bug-free code.

Raspberry Pi

Build

1. Connect the Raspberry Pi exactly like the door setup in Mission 8.
2. Make sure that the crocodile clips are long enough that you don't pull the whole Raspberry Pi out the window when it's opened!
3. Code the Raspberry Pi before you connect it to the window.

Figure 20.13   Raspberry Pi in the window.

Code

Looking closer at the Raspberry Pi code, we can see that this algorithm doesn't apply here. The Raspberry Pi is not getting the inside temperature of the house. It gets the outside temperature according to a website. Our algorithm needs to be different. We could use this as a warning—it's less than 10 degrees Celsius outside; close the window! Or we could look at the other data from the API. Is it raining? Close the window! Let's do both.

1. From the top:
   a. Import all your libraries.
   b. Add your variables for the temperature code from Mission 18.
   c. Set up your window just like you did for the door in Mission 8.
2. Next, I've created some functions to check whether it's raining or if it's cold outside. I've also included the *soundTheAlarm* function from previous missions.

```
1  from urllib.request import urlopen
2  import json
3  import time
4  from gpiozero import Button
5  import os
6
7  #password and map coordinates for temperature
8  apikey="YOURSECRETKEYGOESHERE"
9  lati="51.501401"
10 longi="-0.141895"
11
12 #setup the window
13 window = Button(2)
```

Figure 20.14   Code.

3. Make sure that you've got the siren.wav file (or download your own file). Learn how to do that in Mission 8.

```
1  def isRaining(weather):
2      if weather['currently']['precipProbability'] > 0.5:
3          return True
4      else:
5          return False
6
7  def isCold(weather):
8      if weather['currently']['temperature'] < 10:
9          return True
10     else:
11         return False
12
13 def soundTheAlarm():
14     print ("Close the window!")
15     os.system("aplay siren.wav")
```

Figure 20.15   Code.

4. Get your temperature data, and check whether it's raining or cold outside by calling the preceding functions. Don't forget *time.sleep( )* here. We only want to get the temperature every 10 minutes, or the DarkSky API will be unhappy.

5. If it's raining *or* cold, is the window open? If it is, sound the alarm!

```
1  #get the weather data from the api website
2  url="https://api.forecast.io/forecast/"+apikey+"/"+lati+","+longi+"?units=si"
3
4  while True:
5          time.sleep(600)
6          try:
7              meteo=urlopen(url).read()
8              meteo = meteo.decode('utf-8')
9              weather = json.loads(meteo)
10
11             #is it raining or cold out?
12             if isRaining(weather) or isCold(weather):
13
14                 #is the window open?
15                 if not window.is_pressed:
16                     soundTheAlarm()
17
18         except IOError:
19             print ("Your Internet connection is not working!")
20
```

Figure 20.16   Code.

FUNCTIONS

Finally, we have a full example of how functions work! In the functions *isCold* and *isRaining*, we send them a variable *weather.* The functions do their thing and return a value: *True* or *False*. In the main code, we just call the function in an *if* statement as if it were a variable: *if isRaining(weather) or isCold(weather)*. After the functions return their value, this code will run like this (if it's raining but not cold outside): *if True or False*. Functions make the code easier to read. In fact, *if isRaining(weather) or isCold(weather)* is practically a sentence in English!

6. Add this code to the crontab so that it runs when the Raspberry Pi starts up. Check out Mission 8 on how to do this. Unplug the Raspberry Pi from your monitor, keyboard, and mouse, and set it up attached to the window.

Debug

My debugging on this mission was with the window—getting the tinfoil lined up and touching when the window closed was a bit tricky. A cool debug you could do is to find somewhere in the world where it's raining! Put the coordinates for that place in your code, and see if your alarm goes off.

Expert Level

Nope. Nothing! There is nothing more I can challenge you with. You are an expert.

Index